The Lighthouse on Skerryvore

The Lighthouse on Skerryvore

THE REMARKABLE STORY OF A VICTORIAN ENGINEER WHO DESIGNED AND BUILT 'THE MOST GRACEFUL LIGHTHOUSE IN THE WORLD'

Paul A. Lynn

Whittles Publishing

Published by
Whittles Publishing Ltd.,
Dunbeath,
Caithness, KW6 6EG,
Scotland, UK

www.whittlespublishing.com

ISBN 978-184995-140-1

Printed by

Latimer Trend & Company Ltd., Plymouth

Contents

Preface

The story of the Skerryvore lighthouse and its creator Alan Stevenson is remarkable by any standards. Here was a man of exceptional intelligence and wide-ranging ability, who overcame Herculean challenges over a six-year period to place a lighthouse on an isolated rock in the wild North Atlantic, 12 miles off the Hebridean island of Tiree. But he was also a man of exceptional modesty, often plagued by poor health and self-doubt. By contrast his father, Robert Stevenson, founder of a family dynasty of Scottish lighthouse engineers and builder of the world-famous Bell Rock lighthouse in the North Sea, was an extremely robust character who enjoyed his fame and the Edinburgh lifestyle that went with it. The tension between a father and his eldest son forms an enduring backdrop to the adventure that was Skerryvore.

I became interested in Alan Stevenson some years ago but soon realised that his name and achievements were hardly known to the general public, even in Scotland. His account of Skerryvore was published in 1848, four years after the lighthouse was completed, but it gained nothing like the celebrity of his father's account of the Bell Rock. I managed to purchase a rare first-edition copy, which convinced me that there was a story waiting to be told. My background as a professional engineer has encouraged me to try and enter Alan's world directly, without preconceptions, asking what the state of scientific knowledge and lighthouse engineering was at the time, and what professional and personal hurdles he faced when building 'the most elegant lighthouse in the world'. As a result I have often felt myself in conversation with one of the great engineers of the Victorian era, a largely unsung hero. It has been a fascinating journey.

As a relative newcomer to the delights of pharology, I accept that my choice of material and topics surrounding the Skerryvore story is unlikely to satisfy everyone. Any errors and omissions are certainly not those of the authors and historians whose efforts have lightened my task. I simply hope that my focus on Alan Stevenson's *Account of the Skerryvore Lighthouse* will help bring his remarkable contribution to saving lives at sea to a wider audience.

Paul A. Lynn
Butcombe, Bristol BS40 7XF, UK
paulalynn346@btinternet.com

Acknowledgements and references

I am very grateful to the following copyright owners for permission to use the photographs and images which do so much to bring this story to life:

- **Northern Lighthouse Board, Edinburgh (www.nlb.org.uk):** photos of the Noss Head and Sanda lighthouses (page 112).
- **Ian Cowe, photographer (www.flickr.com/photos/iancowe)**: photos of the Skerryvore lighthouse (front cover and pages 52, 110 and 112) and Bell Rock lighthouse (page 38).
- **Jim Murdoch, Images of Tiree (www.tireeimages.com)**: photos of the Isle of Tiree (pages 40, 41 and 105).
- **Professor Roland Paxton**: image of the Stevenson family (page 122).

All the above are individually acknowledged where they appear in the text, as are a number of Wikipedia images. Additional illustrations have been provided by David Thompson (www.davidthompsonillustration.com) who worked with me on several previous books. It has been a pleasure to repeat the collaboration.

All other images are my own, or are reproduced from historical works that are out of copyright. The Northern Lighthouses insignia which appears on the title page and epilogue is photographed from the front cover of my copy of Alan Stevenson's 1848 *Account of the Skerryvore Lighthouse*. The images of the Bell Rock lighthouse are obtained from Robert Stevenson's 1824 *Account*, digitised by the Internet Archive in 2012 with funding from the Northern Lighthouse Heritage Trust.

I have referred to various other books and websites while writing this book. I make no claims for originality in the historical material on other lighthouses or the Stevenson family, and am grateful for the efforts of those who have researched these topics before me. I would particularly like to mention Bella Bathurst's book *The Lighthouse Stevensons*, Professor Roland Paxton's books *Dynasty of Engineers* and *Bright Lights* (co-authored with Jean Leslie), the Northern Lighthouse Board's website (www.nlb.org.uk), and the reference website for the Bell Rock lighthouse (www.bellrock.org.uk).

Prologue

A wavelet, soon to become a wave, is born in mid-Atlantic. Urged on by the gathering storm, bolstered and encouraged by its neighbours but unaware of its final destination, it heads east across a vast stretch of ocean towards the Scottish Hebrides. How about competing with the hundred-year wave that strikes fear into the heart of mariners and creating havoc on a distant shoreline?

This is no 21st-century tale. We are returning to the 19th: ocean-going vessels powered by the wind; charts approximate or non-existent; no modern aids to navigation; no lifejackets with a hope in hell of holding heads above a raging sea.

The wave, pumped up to the height of a house and speed of a racehorse by three days and nights of unrelenting gale, approaches the isles of Tiree and Iona, with Mull's towering Munro in the background. But something vicious stands in the way, an isolated reef that has threatened sailors for generations and splintered many a merchantman and fishing boat. It is Skerryvore, the year 1840. An unlikely-looking wood and iron construction on stilts, set into the only flattish bit of rock on offer, braces itself for another drenching impact. Thirty men shiver at the top while below, in a space far smaller than a modern prison cell, the engineer distracts himself with notebooks, plans, and high hopes for his new lighthouse, perhaps the most dangerously ambitious project ever undertaken off a European coastline.

An unlikely wood and iron construction on stilts.

But beware of easy assumptions, for Alan Stevenson, son of an illustrious father and uncle of a world-famous storyteller, is no explorer or military man. The spikes of Skerryvore seem ill-suited to one familiar with the drawing rooms of Edinburgh, the novels of Sir Walter Scott and the poetry of William Wordsworth. What combination of upbringing, talent, and determination can possibly have brought him here?

Part 1: Precedents

To the Hebrides

General synopsis at midnight GMT:

New low expected Southeast Iceland …
Viking … Humber, Thames … Finisterre, Sole …
Rockall, Malin, Hebrides:
Northwesterly 7 to gale 8, occasionally severe gale 9.
Rough. Showers. Good, occasionally moderate

As a teenager I often found myself listening, initially by accident but increasingly by design, to the shipping forecast broadcast just before the BBC evening news. Sea area Viking I could cope with – no doubt it had something to do with previously unwelcome Norsemen. Humber and Thames were definitely off England's east coast, but Finisterre and Sole had me floundering. And then, towards the end, came Rockall, Malin, Hebrides, and I realised that the announcer's rhythmic tones had worked us clockwise round the coast of Britain towards the islands and ragged coastline of western Scotland. All this exerted a powerful hold on my imagination as I experienced, by proxy, gales and surging waves in the North Atlantic.

A few years later a group of students, Sassenachs all, coaxed an elderly Morris 8 from London via Perth to Skye, a Hebridean island whose magic still haunts me. My wife and I met there and we have since ranged widely, from Islay and Jura in the south to Harris and Lewis in the north, hill-walking, camping, tempting trout in isolated lochs, and island-hopping on the ubiquitous CalMac ferries.

In the early days I was more or less unaware of lighthouses. Of course, like every English schoolboy I had heard of Grace Darling, the lighthouse keeper's daughter who in 1838 helped her father rescue shipwreck survivors in appalling weather and row them back to the safety of the Longstone lighthouse. Later I chanced upon several fine lighthouses in the west of Scotland, from the Rinns of Islay in the south to Stoer Head in the far north, but had no idea who had designed and built them, or when. Back in England I visited the tower of John Smeaton's famous Eddystone Rock lighthouse, first lit in 1759 and subsequently rebuilt for tourist inspection on Plymouth Hoe, and on one occasion encouraged our young family up a dizzy staircase in the Old Light on Lundy Island. Otherwise, ignorance prevailed.

Then a few years ago a friend suggested Bella Bathurst's book *The Lighthouse Stevensons* as a thoroughly rewarding account of the early Scottish lighthouses and their builders. I was introduced to the Northern Lighthouse Board and its first Engineer, Thomas Smith, who built 11 lighthouses in the period 1787–1806, and Robert Stevenson, Engineer from 1808 to 1842, who added 14 of his own including the iconic Bell Rock off the east coast of Scotland. I learned about his eldest son Alan – the principal hero of this book – and the remarkable Stevenson family dynasty that for well over a century continued to design and build one of the finest historical collections of lighthouses in the world. All told, the Stevenson engineers accounted for around 200 lights and transformed the survival chances of mariners risking their lives in Scottish waters.

Scottish lighthouses built by Thomas Smith and Robert Stevenson, plus the location of Alan Stevenson's Skerryvore near the Isle of Tiree.

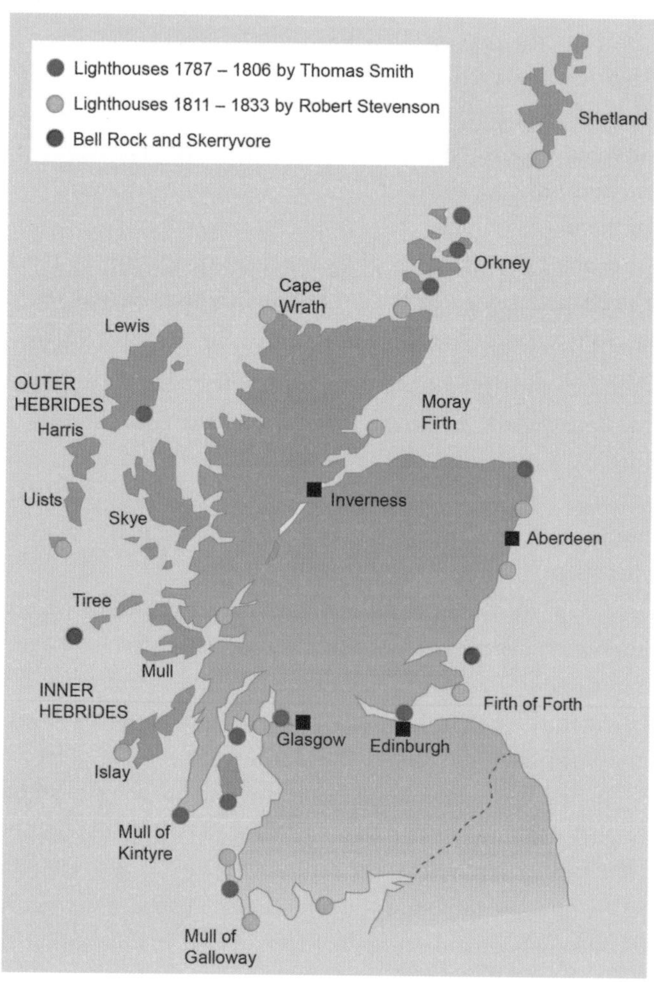

Robert Stevenson's masterpiece, the wave-battered Bell Rock lighthouse that still clings to its isolated reef in the North Sea, had a huge impact on 19th-century lighthouse engineering and greatly influenced the three sons – Alan, David, and Thomas – who followed him into the family business. As Bella Bathurst's story unfolded I found myself increasingly drawn towards the character and achievements of Alan, who succeeded his father as Engineer to the Northern Lighthouse Board in 1843 and built 13 further lights.

Alan's name may not be well known to the general public, even in Scotland, but his story is remarkable in human as well as technical terms, above all for his most audacious creation – Skerryvore, the tallest lighthouse ever built in Scotland, perched on a vicious reef 12 miles out in the North Atlantic from the small Hebridean island of Tiree. Not only is Skerryvore assaulted by exceptional storms but, perhaps unsurprisingly given its designer's many talents, it is sometimes described as the world's most graceful lighthouse.

Alan's father voyaged to the Hebrides on numerous occasions, mainly because one of his principal duties as Engineer was to carry out annual inspections of all the Board's lighthouses, starting from Edinburgh and progressing anticlockwise round the Scottish coast. By the end of his tenure six lighthouses had been built on the far west coast and Hebridean islands: Cape Wrath at the northwest tip of the Scottish mainland; Eilean Glas (Isle of Harris); Barra Head (Isle of Barra); Lismore near Mull; Rinns of Islay (Isle of Islay); and Mull of Kintyre in the south west. The inspections were not confined to technical matters but ranged widely over the duties of lighthouse keepers and the standards expected of them, with shoddy habits and idle housekeeping given extremely short shrift.

Robert Stevenson visited the notorious Skerryvore reef in 1804 and 1814 to assess the chances of crowning it with a lighthouse. He was far from optimistic. On the second occasion, spurred on by his triumphant completion of the Bell Rock three years earlier, he was accompanied by a pleasure-seeking Sir Walter Scott who noted in his diary:

> Pull through a very heavy swell with great difficulty, and approach a tremendous surf dashing over black pointed rocks. Our rowers, however, get the boat into a quiet creek between two rocks, where we contrive to land well wetted. I saw nothing remarkable in my way excepting several seals, which we might have shot, but, in the doubtful circumstances of the landing, we did not care to bring guns. We took possession of the rock in name of the Commissioners, and generously bestowed our own great names on its crags and creeks. The rock was carefully measured by Mr S. It will be a most desolate position for a Lighthouse – the Bell Rock and Eddystone a joke to it, for the nearest land is the wild island of Tyree, at fourteen miles distance. So much for the Skerry Vhor.

Scott's dim view of Skerryvore was shared by the Commissioners, who were fully aware of its dangers and the desirability of lighting it but spent the best part of 30 years procrastinating. Even the passing of an Act of Parliament in 1814 to sanction the project failed to produce action, and it was another 20 years before they sent Robert back to make a detailed survey of the rock. This time he took 27-year-old Alan with him, and reported:

> The rock on which it is proposed to build this Light House forms the foreland of an extensive track of foul ground lying off the Coast of Argyllshire. This reef has long been the terror of the Mariner, but the Erection of a Light House upon Skerrivore would at once change its

> Character and render it a rallying point of the Shipping which frequents these seas.

Terror it may have been, but since there were no reliable charts of the seas around the Hebrides it was almost impossible for mariners to assess where they were and the risks they were taking. The master of one vessel was so unaware of his proximity to the reef that he was found 'lying at ease on the companion, enjoying his pipe, with his wife beside him knitting stockings'.

Alan was acutely aware of the Skerryvore problem long before he took over from his father as Engineer to the Northern Lighthouse Board at the age of 36. Prior to Skerryvore, the tally of Scottish lighthouses stood at about 30, but only one of them, the Bell Rock, was wave-washed – and it stood in the comparatively undemanding waters of the North Sea. Alan had grown up in the knowledge, and probably the shadow, of his father's triumph; the challenge of designing a lighthouse for an even remoter rock environment, long considered a practical impossibility, must have both tempted and frightened him. His own view was certainly realistic:

> From the great difficulty of access to the inhospitable rock of Skerryvore, which is exposed to the full fury of the Atlantic, and is surrounded by an almost perpetual surf, the erection of a Light Tower on its small and rugged surface has always been regarded as an undertaking of the most formidable kind.

Yet he pressed ahead with plans and increasingly frequent visits to the Hebrides, culminating in six years of sustained effort against stupendous odds. Skerryvore, his very own masterpiece, was lit in 1844.

I returned to Bella Bathurst's book recently and became even more hooked on Alan Stevenson and Skerryvore. I also discovered his father's *Account of the Bell Rock lighthouse*, written for the Commissioners of the Northern Lighthouse Board in 1824. And finally, I splashed out on a first-edition copy of Alan's *Account of the Skerryvore Lighthouse with notes on the illumination of Lighthouses,* published in 1848. This rare and wonderful book, full of insights into the mindset of a Victorian engineer, detailed in its account of the technical and human challenges of Skerryvore and adorned with an evocative set of illustrations, has carried me back yet again to Scotland – but this time to the Hebrides of the 1840s.

The Stevensons

Robert Stevenson, founder of the family dynasty that would illuminate the coasts of Scotland, was born in Glasgow in 1772 and had what can only be described as an unsettled childhood. His mother Jean Lillie seems to have eloped to marry his father (another Alan) sometime in 1771 when both were only 20. Tragedy ensured that Robert was to be their only child, for Jean's husband died of the mysterious 'night dews' two years later while on a business trip to the West Indies, leaving her and the toddler to fend for themselves.

Jean's next partner in the marriage stakes was James Hogg, a Glasgow merchant and manufacturer with somewhat dubious business interests. She produced two more sons but he ran off to England with them, leaving her and little Robert in straightened circumstances once again. Determined to try and improve the six-year-old's chances, and remembering her own Edinburgh education, Jean moved back to the capital city and enrolled him in a charity school. Her ambition for him to become a minister in the Church of Scotland was thwarted by a continuing lack of funds so that, by the age of 14, he found himself apprenticed not to God but to an Edinburgh gunsmith.

Robert's turbulent early years began to settle when his mother attended church in Edinburgh's New Town and met the man who would become her third husband. Thomas Smith, born in 1752, had grown up in a village just east of Dundee on the Firth of Tay and suffered early tragedy when his father drowned in Dundee harbour. Accepting his devastated mother's advice that whatever trade he followed it had better be well away from water, he spent five years apprenticed to a Dundee metal worker before moving to Edinburgh and starting his own business manufacturing grates, lamps and metal paraphernalia for New Town's expanding population.

Thomas's combination of talent and ambition proved highly successful in business but his personal life continued tragically. Infectious diseases were terrible tools of the grim reaper in the late 18th century and his first wife, a farmer's daughter, died of whooping cough soon after producing five children, three of whom died in infancy. Thomas remarried, gaining another daughter, but soon lost his second wife to consumption. Jean, who had known and befriended both his wives, moved into Thomas's household to help look after his young family and it is not hard to imagine that she was a willing match for a new engagement. The two grew closer but before they could unite Jean had to track down her second husband, the errant James Hogg. A divorce was eventually procured and she married Thomas in 1787, producing a daughter Elizabeth, known as Betsy, to whom Robert became greatly attached.

Robert had lost contact with his first stepfather and two stepbrothers many years before, and his mother's disastrous second marriage did not even merit a mention in the family bible. Although he appeared as a witness in the divorce proceedings he never told his own children or descendants about it, presumably because he wanted to protect her from scandal.

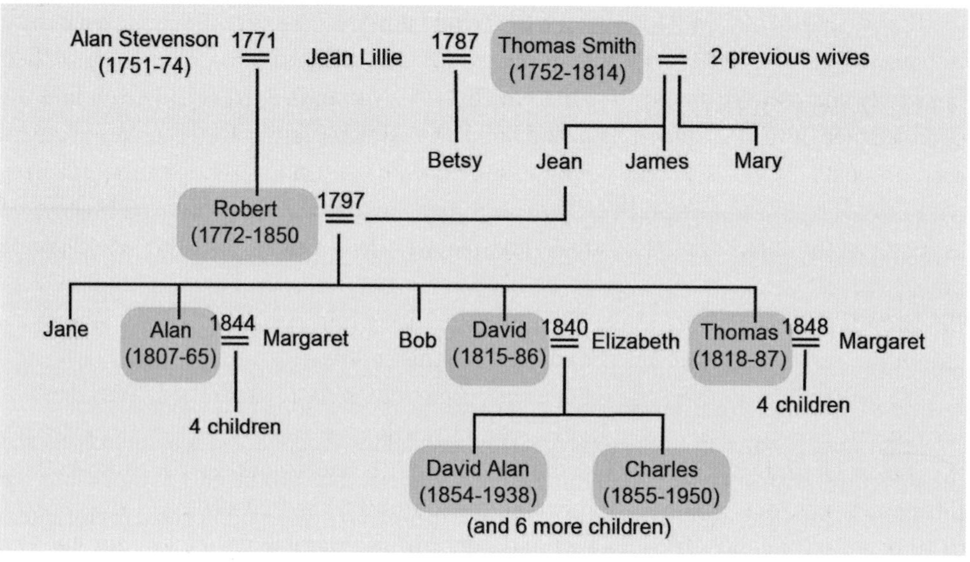

Family connections: Smiths, Stevensons, and the lighthouse engineers.

So by the age of 15 Robert Stevenson found himself the son of a thrice-married mother and the stepson of a man who had lost two previous wives. What he did not yet know was that he would strengthen the Smith–Stevenson family network even further in 1797 by marrying another Jean, the daughter of Thomas Smith by his first marriage. At this point Robert's stepfather also became his father-in-law and so Robert had one Jean for a mother and another for a wife. Jean the younger, also known as Jane, bore him nine children of whom four died in infancy, and three – Alan, David and Thomas – became famous lighthouse engineers. David did his bit to keep the family tradition going well into the 20th century by fathering two sons, David Alan (often referred to as David A) and Charles, who built about 90 further Scottish lights between them.

Stevenson engineers: Robert; his sons Alan, David, and Thomas (Wikipedia).

The makeup of the Smith–Stevenson household must have seemed strange, even quirky, to many onlookers, but it worked. Both halves of the family adopted Edinburgh as their home city and New Town for their lifestyle. By 1803 Thomas's success had given him the confidence to purchase a plot of land and build a fashionable family house, large enough to accommodate the whole tribe in comfort and provide workshop space for experimenting with new designs. As the years went by and Robert matured he was increasingly accepted as head of the family and his strong-minded wife Jean ran a well-ordered Christian household. He would often be away from home on lighthouse business, content to leave the domestic arrangements in his wife's care and accepting her judgement in all matters except the health and especially the education of his children.

The amalgamation of Smith and Stevenson interests not only stabilised Robert's home life but opened up career possibilities beyond his wildest dreams or expectations. He left the gunsmith and was apprenticed to Thomas Smith's metalworking and lamp-making business, which was expanding to include the illumination of lighthouses using a novel form of reflector lamp. His stepfather and father-in-law became his mentor and the two of them never looked back as Robert increasingly helped construct a superb series of pioneering lighthouses. He had discovered his vocation.

Thomas Smith's ingenuity in producing a new design of oil-fired reflector lamp for lighthouses, a huge advance on hopelessly unreliable and inefficient coal fires and candles, was rewarded in 1787 by his appointment as first Engineer to the Northern Lighthouse Board, formed the previous year to light the coasts of Scotland and the Isle of Man. During the following summers Robert helped Thomas with reflector installations and building maintenance, watching and learning as the first lighthouses were designed, built, and commissioned. From 1791 onwards, at what nowadays seems the tender age of 19, he was given increasing responsibility for lighthouse construction.

As Robert developed his practical knowledge of lighthouse engineering he received plenty of encouragement from Thomas Smith to further his scientific education. During the winters of 1792–4 he attended classes in natural philosophy at Glasgow University given by Professor John Anderson, a former soldier with interests in field artillery and experimental physics. Anderson, delightfully nicknamed Jolly Jack Phosphorus by his students, was a key influence in steering Robert towards an engineering career, which he furthered by attending classes at Edinburgh University during the winters of 1800–04. Mathematics, natural philosophy, chemistry, and natural history were on the menu, but his limited knowledge of Latin and total ignorance of Greek prevented him from gaining a degree.

From 1797 onwards Robert was given even more responsibility for constructing Thomas Smith's final three lighthouses: Cloch (Firth of Clyde), Inchkeith (near Edinburgh), and Start Point (Isle of Sanday, Orkney). He became a full partner with Smith in 1802, and when the latter retired as Engineer to the Board in 1808 his own term in office began. He was to complete 14 Scottish lights of his own, starting with Bell Rock (1811) and including two in the Hebrides – Rinns of Islay (1825) and Barra Head (1833).

The breadth of experience gained by Robert as a young man was extraordinary; summers spent travelling round the coast of Scotland, often to extremely remote locations and in

Three of Robert Stevenson's classic lighthouses: Bell Rock, Rinns of Islay, and Barra Head (Wikipedia).

considerable danger, winters back in Edinburgh, studying, experimenting, and planning. Thomas Smith continued with generous encouragement and support, introducing him to the Commissioners of the Northern Lighthouse Board and keenly anticipating the possibility that he would become their next Engineer. All this responsibility, willingly shouldered, meant that by his late teens Robert was a mature adult, still slightly uneasy with books and cultural activities but completely at home with practical engineering in stone, iron, and wood. He came across as a robust and weathered individual with a lively sparkle in his eyes, as tough but fair with others as he was with himself, highly ambitious and determined to make a significant contribution to society. His mother's love and the many sacrifices she had made for him in childhood, the clear evidence from Thomas Smith that talent and hard work brought their rewards, and his personal suitability for a profession combining strenuous physical and mental activity, combined to make him a formidable character. By the time he was appointed Engineer to the Northern Lighthouse Board at the age of 36, he was already engaged on his most famous project – the Bell Rock.

Robert and Jean Stevenson produced nine children between 1801 and 1818, of whom five – Jane, Alan, Bob, David and Thomas – survived. The loss of four infants left Jean fearing greatly for the survivors and she lavished obsessive attention on them. Alan, the eldest son and principal focus of this story, was a pale and rather delicate child who worried her greatly and battled with indifferent health throughout his life. David and Thomas, both to become Engineers to the Northern Lighthouse Board in their turn, were rather more robust in childhood but suffered increasing ill health as they approached retirement. One of the most striking features of the second generation of Stevenson engineers is the contrast between the huge contribution they made to the public good and their private vulnerability. In many ways they led double lives: physically demanding and necessarily unfussy away from home, often on dangerous seas or in remote locations, contrasting with a tendency to hypochondria when back in Edinburgh.

In addition to his elegant masterpiece, Skerryvore, Alan designed 12 Scottish lighthouses including a famous tower in the Egyptian style, Ardnamurchan (1849), which discourages ships from dashing themselves against the most westerly rocks of the Scottish mainland. He retired from his post as Engineer to the Northern Lighthouse Board in 1853 at the age of 46, already suffering from severe ill health, and died 12 years later.

Bella Bathurst's book also covers the lives of David and Thomas, Alan's younger brothers, who followed him and worked in partnership to produce about 30 lights between 1854 and

1880. In the next generation two sons of David, David Alan (David A) and Charles, added a further 90 or so. Charles was to buck the family tendency to fade early, dying in 1950 at the ripe old age of 96. His own son D. Alan Stevenson (1891–1971), the last of the family of engineers, marked the end of an extraordinary contribution to saving lives at sea.

Apart from his engineering work Thomas Stevenson, generally known as Tom, has an unusual claim to fame as father of one of the world's best-loved storytellers. His son Robert Louis Stevenson (1850–94) was initially forced into the family business. However at the age of 21, after some damp and miserable experiences around the coasts of Scotland, Robert Louis announced that he was sick of lighthouse engineering. He put pen to paper instead, and went on to write *Treasure Island* in 1883 and *Kidnapped* in 1886, the latter featuring a shipwreck on a dangerous reef his father had lit 14 years before. The author's relationship with Stevenson engineering remained troubled and ambiguous throughout his short life, a heady cocktail of resentment and guilt, although in his later years he came to appreciate and acknowledge what his grandfather, uncles, and father had achieved:

> There is scarce a deep sea light from the Isle of Man to North Berwick,
> But one of my blood designed it.
> The Bell Rock stands monument for my grandfather;
> The Skerry Vhor for my uncle Alan;
> And when the lights come out along the shores of Scotland,
> I am proud to think that they burn more brightly for the genius of my father.

The Stevenson lighthouse engineers reached the pinnacle of 19th-century endeavour but with the possible exception of Robert, who was blessed with a formidable constitution, many of their achievements were accompanied by more than a fair share of pain and anguish. This is what makes Alan's personal story so compelling.

Sea peril

We do not expect ships to lose their way, founder in rough seas, or strike uncharted reefs. Although commercial fishing remains one of the most dangerous occupations and we are occasionally shocked to learn that a trawler has gone down with all hands, most modern disasters are caused by human error rather than inadequate ships or navigation aids. For example, on a dark night in March 1987 the car ferry MS *Herald of Free Enterprise* gulped a lethal quantity of seawater and capsized shortly after leaving the Belgian port of Zeebrugge bound for Dover, with the loss of 193 passengers and crew. Although the ship's roll-on roll-off design came in for criticism during the endless legal wrangling that followed, the main culprit was found to be the assistant boatswain who was asleep in his cabin when he should have been closing the main bow door. In 2012 the huge Italian cruise ship *Costa Concordia* ran aground on the Isle of Giglio off the coast of Tuscany and capsized with the loss of 32 lives – no fault with the ship; the captain had decided to entertain the islanders with a near-shore salute. Machismo, not machinery.

On the whole we have come to expect modern ships, charts, radio, and satellite navigation aids to give us first-class protection. But seafaring in earlier centuries, especially in rough northern waters close to ragged coastlines, was incredibly dangerous. From the late 1700s onwards the quantity of shipping in Britain's coastal waters increased hugely as the industrial revolution and the flourishing trade it stimulated gathered pace – and there was a corresponding increase in shipwrecks and fatalities, with many hundreds of vessels lost each year around the British Isles. The Scottish situation was particularly distressing because its coastline, studded with islands and semi-submerged reefs, forced captains to choose between offshore passages without shelter and near-shore ones where the rocks were waiting for them. As Robert Stevenson put it:

> The danger of falling in prematurely with the land during the night, and the rapidity of the tides on these shores, induced the mariner to keep along the extreme points and headlands of the coast, holding his course even to the northward of Orkney and Shetland, and to the westward of the Lewis Isles by St Kilda, exposed to the heavy seas of the Atlantic Ocean. In this way, much hazard to shipping, and loss of time, were incurred; and when overtaken with gales of wind, such vessels were unable to avail themselves of the numerous bays and anchorages of the Highlands; – considerations of much importance to heavy laden ships, but especially to the smaller classes of coasting and fishing vessels. It therefore appeared, that nothing but the erection of Lighthouses, by which the mariner might identify the land under night, would render this navigation at all a safe one.

By the 1780s the situation had become so serious, and public concern and agitation so strong, that Parliament sanctioned the construction of four lighthouses at key points around the Scottish coast. In 1786 the Northern Lighthouse Board was born and the rest, as we know, is history – as pioneered by Thomas Smith and Robert Stevenson.

Yet nearly 60 years would pass before Alan made his mark on Skerryvore. Understandably, the Board's Commissioners focussed initially on lighthouses that could be erected without the technical difficulty, expense, and human danger posed by isolated wave-washed rocks. The Hebrides were still sparsely illuminated and the extensive reef containing Skerryvore remained a particular hazard. In his *Account of the Skerryvore Lighthouse,* Alan gives a poignant reminder:

> That this source of danger to shipping was by no means imaginary, and the consequent terror of mariners far from being ill-founded, there is a too melancholy proof in the following list of disasters caused by the Skerryvore Rock, and the neighbouring dangers off the coast of Tyree:

He then lists 31 vessels that the Reverend Neil Maclean, Minister of Tiree and the neighbouring island of Coll, advised him had been lost within sight of Skerryvore between 1790 and 1844, many carrying the names of maidens or messages of hope that, down the ages, have cushioned sailors against the dread of shipwreck. Among them:

> In 1790. The Ship Rebecca of 700 tons lost; crew saved.
>
> ...
>
> 1817. A Brig, of 400 tons, foundered off Kennavarah, Tyree; crew all drowned. Numerous casks of butter came ashore.
>
> ...
>
> 1828. Sloop, Delight, of 70 tons, Stevenson, Master, lost.
>
> ...
>
> 1841. April 2. Majestic of North Shields, Tait, master, of 400 tons, foundered by a sea off Boinshly Rock, and came ashore at Gott Bay; captain and four men washed overboard and drowned, and the mate and one seaman had their legs broken when the vessel was struck by the sea.
>
> ...
>
> 1842. A Barra Boat wrecked, and four corpses washed ashore; two men, a woman, and a child.

Unfortunately the reverend's melancholy list was far from complete. Alan notes that many vessels went down without trace; others left grim calling cards of driftwood, cargo, or bodies on local shores – but rarely on Skerryvore itself because the stormy seas tended to scour it clean. Identification was often impossible and many foreign ships tersely recorded as 'lost at sea' had undoubtedly met their fate on the dreaded reef.

However we should not assume that everyone was in favour of lighthouses. A few religious zealots proclaimed that if God wished to save lives at sea he could certainly do so, and that the loss of a ship was a form of divine retribution. Less extreme but more common was a view widely held in impoverished coastal communities, that a wreck which deposited driftwood and cargo ashore was a blessing to be welcomed. In some areas of Britain – Cornwall is the best known – it was widely rumoured that ships were actively enticed onto the rocks by parties of wreckers showing false lights who had no qualms about dispatching survivors and stealing the booty. In the Hebridean islands landlords were said to charge tenants higher rents if they were lucky enough to face a bountiful ocean. All these factors and attitudes were well known to Thomas Smith and, in their turn, the Stevensons.

Most large vessels plying Hebridean waters in the mid-19th century were traditional sailing ships powered by the wind. Typically they were sloops of 50 to 70 tons; two-masted brigs, naval as well as commercial, of a few hundred tons; and a range of three-masted vessels including schooners, frigates and barques. The reverend's list did not mention the size of anything under 50 tons and many of the entries, including fishing boats, must have been a great deal smaller. It was the larger wrecks that preoccupied owners and insurers worried about money, clergymen comforting widows, and local islanders awaiting the greater booty of cargo and splintered wood they washed ashore.

However not all large wrecks were wooden sailing ships. The first half of the 19th century – critical years in the history of Scottish lighthouses – saw the steady introduction of iron for shipbuilding and steam power for propulsion. Paddle-steamers were plying between Liverpool and Glasgow by 1815 and just 30 years later Brunel's famous ship, the screw-driven SS *Great Britain*, at 3,700 tons the largest vessel afloat, became the first iron steamer to cross the Atlantic. On her maiden voyage from Liverpool to New York in 1845 she passed within sight of Robert Stevenson's lighthouse on the Mull of Galloway, and barely 50 miles south of Alan's brand new creation, Skerryvore, which had been lit the year before. Iron and steam, as well as wood and sail, were increasingly grateful for the Northern Lighthouse Board and its famous motto *in salutem omnium* ('for the safety of all').

All of which makes the unobserved and unrecorded plight of small wooden boats and their crews the more pitiable. It is hard to imagine the terror of local fishermen in a Hebridean storm, especially when the nearest shelter was as distant as the Isle of Tiree from Skerryvore. One of many evocative illustrations in Alan's *Account of the Skerryvore Lighthouse* strikes a jarring note here, because it shows his graceful and majestic lighthouse set on its rock above an idle ocean. And if we focus on the base of the tower we notice two small, open, sailing boats equipped with the most basic form of square rigging, pottering around as though on a Sunday afternoon picnic. The reality is that the crews, 12 miles from the nearest land, would be in great peril if the sea got up. Surely the artist would have been wiser, and Alan better served, to depict a two-masted brig being tossed about at a safe distance, with the lighthouse successfully resisting all the North Atlantic could throw at it.

'As though on a Sunday afternoon picnic.'

English lesson: Eddystone

Professional engineers rarely start from scratch, preferring to adapt and refine existing designs to take account of experience and technical progress. There are occasional examples of something revolutionary and the great British engineers of the early 19th century produced their fair share – Thomas Telford comes to mind with his Pontcysyllte canal aqueduct (1805) high above the River Dee, and Isambard Kingdom Brunel with his SS *Great Britain* (1843). It would be nice to report that Robert Stevenson conjured his design for the Bell Rock lighthouse out of thin air, but fairer to admit that he was strongly influenced by the work of John Smeaton who in 1759 had built the third lighthouse to stand on one of England's most notorious offshore rocks, the Eddystone. Alan, in turn, was steeped in the history of both Eddystone and Bell Rock by the time he came to confront Skerryvore.

The Eddystone Rock forms part of a reef in the English Channel about 13 miles southwest of Plymouth Sound, one of England's most historic naval harbours. It lurks just below the surface at high spring tides. Over many centuries sailing ships faced a double threat; either they foundered on the reef itself or, giving it a wide berth, risked deadly contact with the northern coast of France.

The first attempt at taming Eddystone was the brainchild of Henry Winstanley, an English eccentric and showman. Born in Essex in 1644, he was a man of many tastes and talents. Inspired by a Grand Tour of Europe he became an expert in architectural engraving. He then created an 'Essex House of Wonders' in his home to demonstrate mechanical and hydraulic gadgets and curiosities, and capped it all by setting up a highly successful public entertainment known as 'Winstanley's Water-Works' in London's Piccadilly. Fancying the idea of becoming a merchant he bought five ships, two of which met their fate on Eddystone. The Admiralty assured him that the rock was impossible to mark with a light and that wrecks were unavoidable, but Winstanley refused to accept defeat and announced that he would build a lighthouse himself. The flamboyant design he came up with suggests that he regarded it as an extension of his earlier gadgets, refashioned on a grand scale. Little did he know what he was in for.

The lighthouse, an octagonal wooden structure, took two years to build and was first lit in 1698. England was at war with France at the time and during construction a French privateer captured Winstanley and transported him to France, prompting Louis XIV to order his release with the words 'France is at war with England, not with humanity'. The finished lighthouse was damaged in its first winter and subsequently modified to a 12-sided stone-clad exterior with an octagonal top section, a glass lantern room powered by candles, and ornamental trimmings which reflected Winstanley's dramatic sensibilities. But derision would be unfair because the lighthouse was the first ever to be erected on a small rock in the open sea and it remained lit for five valuable years. The denouement came in November 1703 when a great storm, the most severe ever recorded in southern

England, destroyed 13 ships of the Royal Navy, piled up hundreds of vessels on the River Thames in London, killed over 1,500 seafarers, demolished the Eddystone lighthouse and took the lives of six men working inside it. Henry Winstanley was one of them. The English eccentric had declared that he would relish being inside during 'the greatest storm there ever was' and continued to express faith in his creation to the last.

Next up for the Eddystone challenge was a Captain Lovett who obtained a 99-year lease on the now-vacant rock and Parliament's permission to charge passing ships a toll of a penny per ton once he had lit it. He commissioned John Rudyerd, a London silk merchant, to design him a lighthouse. It seems extraordinary that Lovett would risk entrusting the project to a complete tyro, but the engineering profession was in its infancy and large and difficult projects were often tackled by wealthy amateurs, adventurers and hobbyists – as Winstanley had already demonstrated. In the event Rudyerd came up with a plausible design for a conical wooden tower, which was erected in 1709 and lasted for a highly creditable 47 years. Its demise was not caused by a storm but by a fire which broke out at the top of the lantern, probably caused by a spark from one of the candles, forcing the keepers down and out onto the rock. The disaster is best remembered in folklore for the tragic demise of 94 year-old Henry Hall, the keeper on watch, who threw water upwards at the fire from a bucket, ingested molten lead coming downwards, and died an agonising 12 days later.

The benefits of lighting the Eddystone rock were by now clear to all and the third lighthouse was not long coming. Its creator John Smeaton (1724–92) marks a change from amateurism towards modern professional engineering. Born in Yorkshire, he was educated at Leeds Grammar School and trained as an instrument maker. Later he was to become involved in a wide range of engineering projects including harbours, bridges and canals, and he was one of the first to call himself a civil engineer. It is hardly surprising that

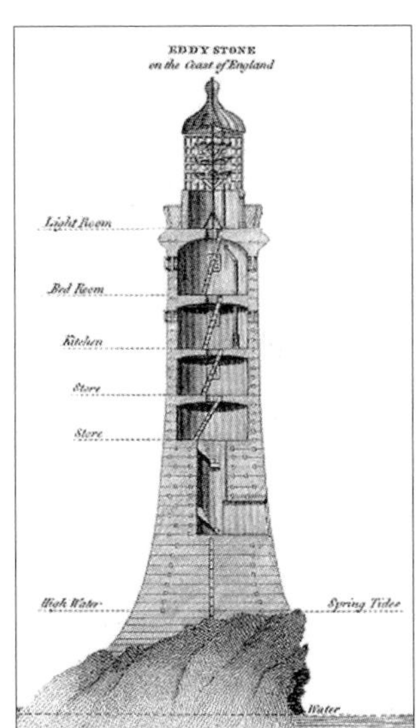

The Eddystone lighthouses of Winstanley and Smeaton. Smeaton's Tower (right) now stands on Plymouth Hoe (Wikipedia).

the Royal Society of London recommended him as designer of the much-needed third lighthouse, which he conceived in 1755 and finished in 1759. It was to prove a highly influential English lesson for Robert Stevenson.

Smeaton's design was revolutionary in several respects. Famously, he modelled the shape of the tower on the trunk of an English oak, considering that its tapering form gave exceptional strength and resilience:

> Why? Because the oak tree resists similar elemental pressures to those which wrecked the [Winstanley] lighthouse; an oak tree is broad at its base, curves inwards at its waist, becomes narrower towards the top. We seldom hear of a mature oak being uprooted.

Rather than work in wood and iron he developed a method of fashioning and securing massive granite blocks with dovetail joints. Each horizontal stone layer, or 'course', was fitted together like a jigsaw puzzle and discouraged from shifting over its neighbours by vertical wooden dowels or 'trenails' and stone 'joggles'. Smeaton also investigated and pioneered the use of 'hydraulic lime', a form of concrete that sets under water. The finished tower contained nearly 1,500 granite blocks and weighed 1,000 tons, with a diameter of 26 feet at the base, tapering to 15 feet at the top and reaching a height of 68 feet. The lantern contained 24 tallow candles and was first lit in 1759. Smeaton's sea-washed wonder guarded the Eddystone for over a century but steady erosion of the rocks beneath the tower caused it to shake menacingly in storm conditions. It was dismantled in 1877 and rebuilt on Plymouth Hoe as a memorial, where it attracts lighthouse enthusiasts and tourists to this day. The foundations and solid granite stub remain out on the rock, as storm-battered as ever. A successor lighthouse, built nearby, was lit in 1882.

Smeaton's creation had already seen 40 years' service by the time Robert Stevenson began to help erect Scottish lighthouses and he was fully aware of the design. In 1801, at the age of 29, he set out on a tour of English lights, visiting 14 and comparing their construction and use of materials with the techniques Thomas Smith had developed in Scotland. Generally, he found the English lighthouses less sturdy and he disliked the way they were built by private individuals in exchange for tolls charged to ship owners. He was disappointed not to reach the wave-washed Eddystone on his first tour, but returned to England in 1813 determined to inspect the lighthouse that had given him so much inspiration and food for thought. As often happens when admirer meets hero in the flesh, he found Smeaton's tower something of a disappointment, partly because it looked 'rather diminutive' and partly because it was not being properly maintained. Five years later he paid another visit, found it in better shape, and gave the underlying rock a careful inspection. He concluded, correctly, that wave erosion was hollowing out the Eddystone rock to a dangerous extent, and must have felt mightily relieved that his own Bell Rock lighthouse was more firmly based.

Scottish lesson: Bell Rock

> SIRE,
>
> It is with much diffidence that the author now lays before Your Majesty, an Account of the arduous national undertaking of erecting a Light-house on the Bell Rock, – a sunk reef, lying about eleven miles from the shore, and so situated as to have long proved an object of dread to mariners on the eastern coast of Scotland, especially when making for the Friths [*sic*] of Forth and Tay.

With these words Robert Stevenson dedicated his *Account of the Bell Rock Light-house* to King George IV in 1824. Whether George's heavy drinking and indulgent lifestyle gave him the taste, or time, to read an amazingly detailed account of a Scottish lighthouse seems doubtful, but it has come down to us as an inspiring record of one man's struggle with rocks and sea 400 miles north of his majesty's shore-based pleasures in Brighton Pavilion. It was essential reading for Alan who, 14 years later, pitted his wits against Skerryvore.

The Bell Rock, also known as the Inchcape Rock and immortalised in a famous poem of 1820 by Robert Southey, forms part of a long and treacherous sandstone reef in the North Sea about 11 miles south-east of Arbroath. It had been a scourge of vessels plying

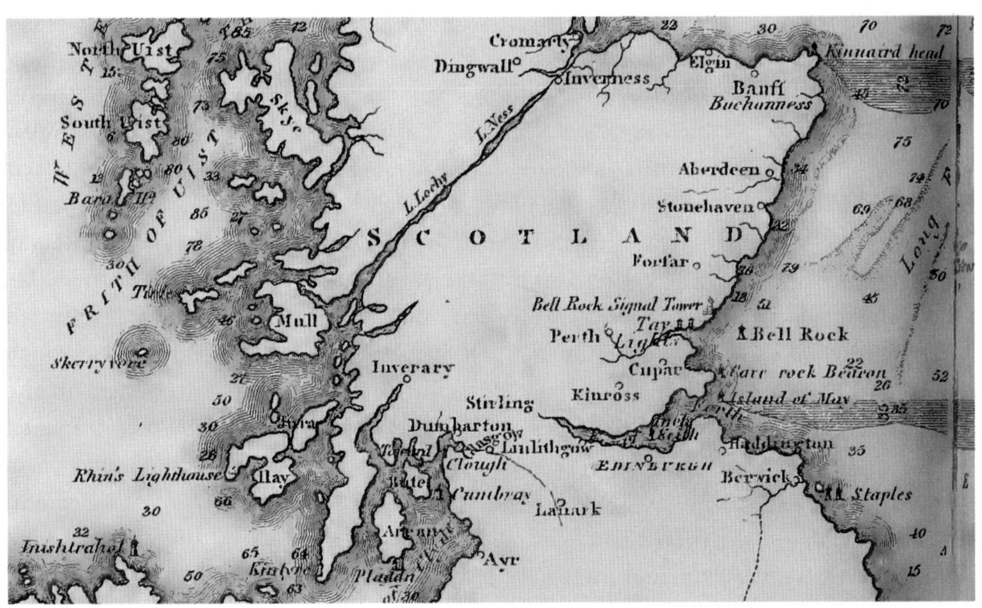

Part of Robert Stevenson's 1824 map of Scotland showing the Bell Rock lighthouse and Edinburgh on the right, and Skerryvore (awaiting Alan Stevenson's attention) on the left. Also marked is the Bell Rock Signal Tower, built some 11 miles from the Rock in the port of Arbroath.

Scotland's east coast for centuries. The reef extends for about 1,500 feet, lurking below the sea surface for much of the time and producing dangerous eddies. According to legend the Abbot of Arbroath placed a bell on the rock in the 14th century, only to have a Dutch pirate remove it. Over hundreds of years the reef remained unmarked and dreaded and by the late 18th century it was exacting a terrible toll on naval and merchant shipping.

From its establishment in 1786 the Northern Lighthouse Board was left in no doubt about the Bell Rock and was under constant pressure to do something about it. Petitions were raised, Parliamentary questions asked, and a hotchpotch of schemes proposed. There were, unsurprisingly, two main problems: the perceived danger and technical difficulty of crowning a wave-washed rock with a lighthouse; and the huge expense. The Commissioners, short of cash, were fully stretched building and maintaining a portfolio of shore-based lights and had no stomach for what many of them considered a madcap scheme. Anyway, Parliamentary permission would be needed for a Bell Rock lighthouse, and this would no doubt delay the project even further.

Yet the pressure continued. These were the years of the Napoleonic wars and the Royal Navy, heavily committed to defending the British Isles as well as attacking French warships on the high seas, could ill afford to lose ships on Scottish rocks. Its voice weighed heavily in the corridors of power, adding to the cries of commercial ship owners, merchants, and philanthropists. In 1803 a Parliamentary bill sanctioning a lighthouse was finally presented to the House of Commons but thrown out by the Lords. And then a seminal event occurred – the loss of HMS York, a three-masted, 1400-ton, 64-gun ship of the line of the Royal Navy, destroyed on the Bell Rock during a routine patrol of the North Sea in January 1804. The entire crew of 491 men and boys perished. It was another turn of the screw.

The Bell Rock menace had already been on Robert Stevenson's mind for a number of years. He was crucially aware of the rock's ability to splinter ships approaching or leaving the Firths of Tay and Forth – the latter hosting Leith, the port of his home city Edinburgh. For several years before the loss of HMS *York* he had been badgering the Commissioners to act and championing himself as the man for the job. He visited the reef in 1800, in a boat crewed by fishermen who were mainly interested in treasure-hunting for items of shipwreck. He surveyed the rock carefully and, having toyed with the idea of a lighthouse raised on iron pillars, firmed up on plans for a stone tower:

> The depth at high-water upon the Bell Rock was much against the design of a building with pillars, as a vessel drawing 12 feet water, and loaded with 100 or even 200 tons, may come with full sail against any erection made there. Were such a circumstance to happen to a pillar-formed building, and a ship to get thus entangled among the openings of the under part of the light-house, there is little doubt that the event would prove fatal to a building of that construction, however strongly framed. On the contrary, supposing a vessel to strike a building of stone, under these circumstances, it is not at all likely, that she could have any effect upon a mass of matter extending to 2000 or 3000 tons, so as to injure such a fabric.

From now on Robert was determined to press for a stone building similar to Smeaton's Eddystone. With a decade of experience erecting lighthouse towers for Thomas Smith, he was riding on a wave of self-confidence, convinced he could make his own special contribution to safety at sea. But the Commissioners had their doubts. True, Robert had been taking increasing responsibility for shore-based Scottish lights, but he was young and his reputation was local. To entrust him with such a perilous and expensive adventure as Bell Rock was more than the Commissioners dared risk – so they took the obvious alternative of seeking an engineer with a national, or even international, reputation, and appointing Robert as his deputy.

There were two obvious candidates and both had the advantage of Scottish birth. The first, Thomas Telford, was already famous for opening up the Highlands with a monumental programme of road and bridge construction. He was just completing his majestic Pontcysyllte aqueduct in North Wales and was busy on the Caledonian Canal. Initially tempted by the Bell Rock project, he submitted a design and estimate for a lighthouse but subsequently decided he was too busy with other projects to continue.

The Commissioners next turned to John Rennie, a man of 44 with an established reputation for building canals, aqueducts, and bridges who was currently working on the London Docks. Rennie had no prior experience of lighthouses but his success in tackling a wide variety of engineering challenges seemed recommendation enough. They decided to appoint him Engineer, with Robert as his assistant. The two men visited the Bell Rock soon after the HMS *York* disaster when, according to Robert:

> They made a favourable landing; and Mr Rennie had only been a short time upon the rock, when he gave his decided opinion upon the practicability of the proposed erection of stone. He had examined the author's designs and models, and afterwards made a Report, in which he coincided with him in recommending to the Board the adoption of a building of stone, on the principles of the Eddystone Light-house.

It all sounds sweetness and light but in fact Robert was deeply troubled. He had already spent years pondering and planning the Bell Rock and had an unshakeable conviction that he was the man for the job. Sensing that his baby was being snatched away, he wrote to Rennie enclosing detailed plans and adding plaintively that the lighthouse was 'a subject which has cost me much, very much, trouble and consideration'. Rennie could hardly accept the role of Engineer without recommending some ideas of his own, and in view of his inexperience it is unsurprising that he suggested modifying Robert's proposal to nudge it closer to Smeaton's Eddystone design. It was the start of a long professional struggle between the two men – which seems, however, not to have ruined their personal relationship. This was probably thanks to Rennie's good nature and the realisation that he had enough on his plate, rather than to a spirit of give and take on the part of the young terrier snapping at his heels. As the work proceeded, Rennie's contribution dwindled to giving advice from a distance – advice that was sometimes ignored – and Robert's part in the enterprise blossomed to such an extent that, in the public mind as well as that of the Northern Lighthouse Board, it became a Stevenson lighthouse. When Robert died in 1850 the Commissioners acknowledged that his was 'the honour of conceiving and executing the great work of the Bell Rock lighthouse'. Rennie's descendants continued to argue the toss, but to little effect.

However expert opinion nowadays tends to give John Rennie considerably more credit for the design of the Bell Rock tower than Robert did in his *Account*. In particular, Rennie recommended flaring the base out more to reduce its resistance to the waves, and insisted on extensive dovetailing as in the Eddystone. It can be argued that his modifications to Robert's initial ideas were key to the tower's long-term survival; but they owed much to Smeaton and were not, in this sense, original. Rennie seems to have visited the site only four times during the five-year build, whereas Robert immersed himself totally in the project, pioneering novel construction techniques and facing great personal danger.

Perhaps it all boils down to semantics. What exactly do we mean by saying that someone 'built' a lighthouse? Is 'conceiving and executing' the same as 'designing and constructing'?

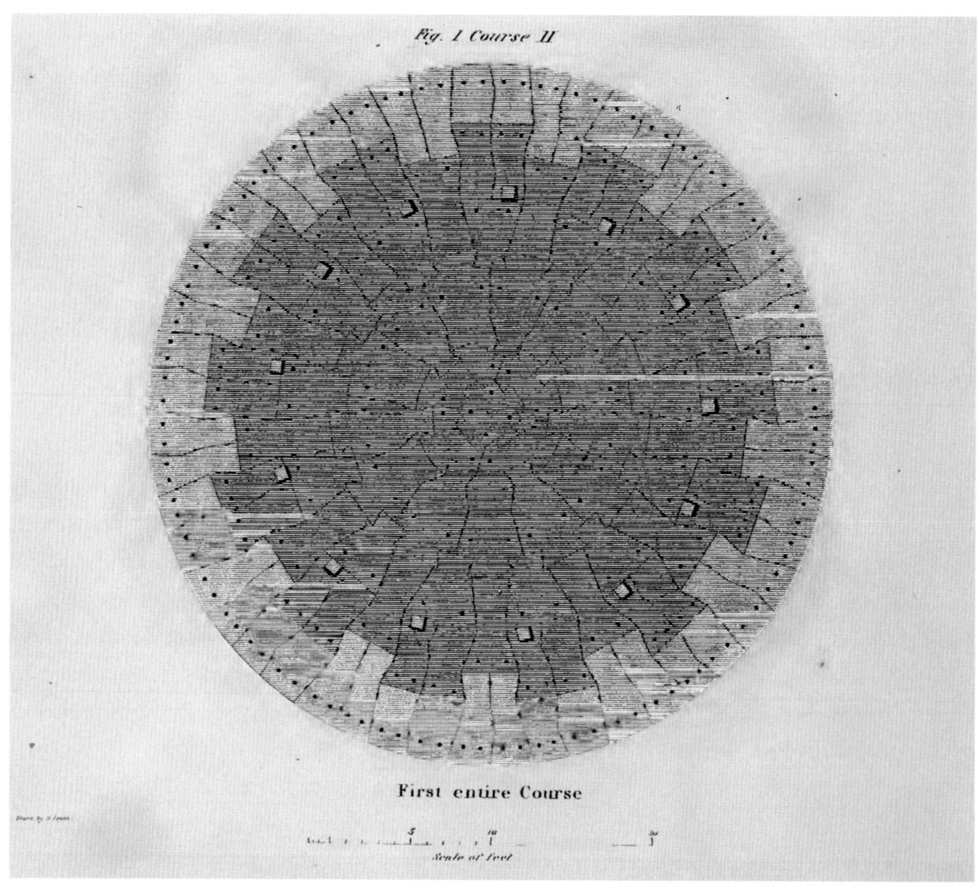

The first entire course of stones as built, showing the trenails (black dots), stone joggles (small squares), and extensive dovetailing.

Rennie clearly made an important, and possibly vital, contribution to the design but it was office-based and he had little to do with construction. Robert was hardly a theorist, but he had spent many years championing a Bell Rock lighthouse and pondering its design before Rennie arrived on the scene. It is hardly surprising – even if less than admirable – that he wanted the lion's share of the credit for himself.

Work on the rock began in the summer of 1807. The team depended crucially on two support vessels: a 67-foot, 82-ton, flat-bottomed craft previously captured from the Prussians and renamed *Pharos*, which was equipped with three large copper lanterns and acted as a Floating Light moored close to the rock; and the 42-ton *Smeaton*, generously named after the hero of Eddystone, a brand new 55-foot supply vessel for ferrying men and materials, including blocks of dressed stone, from the port of Arbroath some 11 miles to the north. Also purchased were two smaller praam boats, *Dickie* and *Fernie*, for transferring men and materials from the *Smeaton* and working close to the rock.

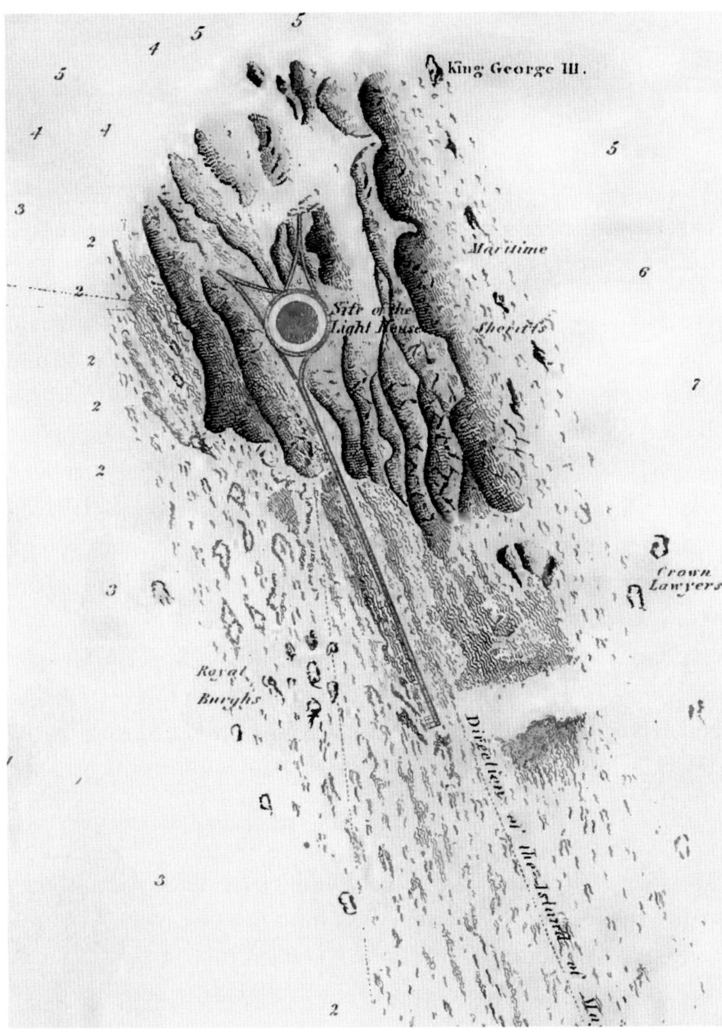

Robert's plan view of the Bell Rock at low tide, showing the site of the lighthouse and the railway tracks used to shift stones from the landing places. As the work proceeded, many of the reef's features were named after distinguished visitors.

Transferring blocks of stone from the Smeaton to a praam boat.

In the early stages the *Smeaton* and *Pharos* were also used to billet the small but growing army of artificers – skilled masons, smiths, millwrights and joiners – who agreed to remain on duty at the rock for a month at a time in exchange for wages of £1 a week plus food and copious quantities of beer. *Smeaton's* accommodation proved woefully cramped and the flat-bottomed *Pharos* wallowed horribly in North Sea swells. There was general relief when the 84-ton schooner *Sir Joseph Banks*, specially built to accommodate the artificers, arrived the following year. But seasickness was a continuing menace, even when partly relieved by munching dulse, an unappetising dark red seaweed; and there was constant fear of accidents when transferring between vessels and rock, especially among landlubbers who dreaded the very thought of such a location.

Robert Stevenson, tough and resolute as he was, did not belittle their fears. On one occasion he shared those fears, almost to despair. The *Smeaton* broke loose from her mooring in a rough sea and drifted three miles downwind, leaving him and 32 men on the rock, threatened by a rising tide. Robert realised that the boat could not possibly get back in time to save them all:

> In this perilous predicament, indeed, he found himself placed between hope and despair,– but certainly the latter was by much the most predominant feeling of his mind,– situate upon a sunken rock in the middle of the ocean, which, in the progress of the flood-tide, was to be laid under water to the depth of at least twelve feet in a stormy sea.

The men, hard at work, had not noticed the developing crisis. But as the tide rose and they prepared to vacate the rock, they became only too aware of their predicament. Robert prepared to address them with instructions and words of encouragement:

> But when he attempted to speak, his mouth was so parched, that his tongue refused utterance, and he now learned by experience that the saliva is as necessary as the tongue itself for speech. He then turned to one of the pools on the rock and lapped a little water, which produced immediate relief. But what was his happiness, when, on rising from this unpleasant beverage, some one called out 'A boat, a boat!'

Neither saliva nor speech was needed, for a boat had appeared unexpectedly through the haze, bringing letters from Arbroath. It was a desperately close shave.

On the rock, the first season's work concentrated on setting up a forge, constructing a Beacon House that would subsequently serve as a temporary barrack for Robert and his workforce, and starting excavation of a foundation pit for the lighthouse tower. Apart from storms and heavy swells which, even in the summer months, frequently interrupted the work, tides were the major problem. Unlike the Eddystone, the Bell Rock spent much of its time submerged at depths of up to 12 feet. During spring tides it was possible to stay on it for several hours around low-water before all work had to be suspended, but during neaps there was often no window of opportunity at all. Even with working days beginning at five in the morning and ending at eight or nine in the evening, much time was necessarily spent waiting, fishing, and chatting. So much so that Robert, challenging the era's strict social conventions, convinced the Board and the artificers that Sunday working was essential. His laboured prose suggests an uneasy conscience:

> Surely, if, under any circumstances, it is allowable to go about the ordinary labours of mankind on Sundays, that of the erection of a light-house upon the Bell Rock, seems to be one of the most pressing calls which could in any case occur, and carries along with it the imperious language of necessity. When we take into consideration, that, in its effects, this work was to operate in a direct manner for the safety of many valuable lives and much property, the beautiful and simple parables of the Holy Scriptures, inculcating works of necessity and mercy, must present themselves to every mind unbiased by the trammels of form or the influence of a distorted imagination.

From our 21st-century viewpoint it is hard to imagine the Herculean labour involved in shifting heavy materials, manoeuvring small boats in heavy seas, transferring blocks of stone onto a dangerous rock, and shaping stone to tight tolerances. There were no power tools; everything had to be done by hand. A good example was boring holes in the rock to take iron fixings for the great timber legs of the Beacon House:

> The operation of boring or drilling these deep holes in the rock, was conducted with great dexterity in the following manner: Three men were attached to each jumper or chisel; one placed himself in a sitting position, to guide the instrument and give it a turn at each blow of the hammer; he also sponged or cleaned out the hole, and supplied it occasionally with a little water; while the other two, with hammers of sixteen pounds weight, struck the jumper alternately, generally bringing the hammer with a swing round the shoulder, after the manner of blacksmith's work .... After many observations, as to the time occupied in boring these holes, the writer found that, when the tools were of very good temper, they could be sunk at the rate of one inch per minute, including stoppages.

By October 1807 Robert must have felt well pleased with progress. The Beacon House was well under way, trial landings of stones from the praam boats onto the rock had been successful, and the foundation pit for the lighthouse was taking shape. As the season drew to a close, John Rennie, accompanied by his son, made a rare visit to inspect progress, advising Robert that his pleasure 'will be heightened if in the interim you can bargain with old Neptune to favour us with a quiet sea while I am on board the Floating Light. I hate your rolling seas'. Fortunately Rennie was impressed, the visit was a success, and the Rennies left the rock fortified by a farewell toast.

Landing stones from a praam boat onto the Bell Rock.

After a winter spent back in Edinburgh, Robert returned to the Bell Rock with an expanding workforce in May 1808. The Beacon House, although far from finished, had survived the winter storms, and the *Sir Joseph Banks* was now on station with accommodation for up to 40 artificers, 15 crew, the engineer and his foremen, and a landing-master. Many building materials and tools had been prepared at the workyard in Arbroath over the winter months, including granite quarried near Aberdeen and sandstone near Dundee. By the end of May 60 artificers were employed at the yard and a month later the Bell Rock itself was teeming with activity:

> The surface of the Rock was crowded with men, the two forges flaming, the one above the other, upon the Beacon, while the anvils thundered with the rebounding noise of their wooden supports, and formed a curious contrast with the occasional clamour of the surges.

The rest of 1808 saw major advances towards Robert's long-held dream. Cast-iron rails were delivered by the *Smeaton* and railways built to transport blocks of stone from landing places to the site of the tower. By the end of the season the foundation pit was completed and the first three full courses of stones were laid.

The foundation pit had proved especially troublesome, not least because it was flooded by each rising tide and had to be pumped out before work could restart. It was also much larger than Smeaton's pit on Eddystone – 42 feet rather than 26 feet in diameter, with 2.6 times the area. The task was huge, especially since the Bell Rock sloped markedly from one side to the other yet the pit had to be perfectly level. As the masons' picks worked deeper they found that the rock grew harder and that some portions were flawed to a greater depth. Robert decided that a 'partial' course of stones would have to be inserted to provide a sound base for the first full course.

The foundation pit. On the left, smiths work the forge; on the right, a team pumps out seawater. Twin forges were subsequently raised on a platform in the Beacon House to prevent the flames being extinguished by the rising tide.

The plan for the lighthouse tower was based on the Eddystone, with one major difference – height. Smeaton had built his tower 68 feet high, but Robert was aiming for a much more impressive 100-footer with a correspondingly greater diameter at the base.

The main reason for building a lighthouse high is to counteract the Earth's curvature and make the light visible at a greater distance. Robert makes little of this in his *Account* and it is far from obvious that the Bell Rock's location demanded a greater range than the Eddystone. Perhaps his decision had more to do with trumping the preceding ace than meeting the strict demands of navigation. Perhaps '100 feet' had a certain ring about it. Whatever the justification, the height increase had major implications for the size of the foundation pit, the number of stones required, the labour of transportation and erection, and the overall cost. The weight of Robert's lighthouse would come out at 2,083 tons, just over twice that of the Eddystone, and its cost of £61,331 would be about 50 per cent greater. One wonders whether the cash-strapped Commissioners of the Northern Lighthouse Board were fully aware of the implications.

On Sunday 10th July 1808 the foundation stone of 20 cubic feet was ready for laying. Never one to miss a chance for celebrating his engineering achievements, Robert led the work and was ready with a suitable benediction:

> 'May the Great Architect of the Universe complete and bless this building', on which three hearty cheers were given, and success to the future operations was drunk with the greatest enthusiasm.

Construction of the tower now proceeded apace. Low tide determined when work could start and in high summer the wake-up call sometimes came as early as 3am. If the landing was to be made before breakfast, the men were first fortified with a wee dram and a biscuit. Although the sea hampered access to the rock in all but the calmest conditions, August saw the first complete course of 123 dovetailed stones laid in Eddystone style. By the end of September 1808, and in spite of delays caused by a shortage of granite and some vicious weather, a solid stub of three courses was in place, submerged by each high tide and ready to face the trials of winter. Robert estimated that 400 blocks with an average weight of nearly 1 ton had been laid during 265 working hours on the rock, and as he returned to Edinburgh he must have felt relieved that the master plan, so long a design on paper, was finally taking shape.

The 1808 season, although highly successful in engineering terms, had given Robert plenty of managerial and organisational headaches. The notorious press gangs were busy around Arbroath and Dundee, seizing unsuspecting merchant seamen and hauling them off to serve in his majesty's warships. Robert's boat crews, in danger of impressment when ashore, had to be issued with protection tickets confirming they were urgently needed on the Bell Rock. There were other dangers. One of Robert's best masons, Hugh Rose, was grievously injured when a 2-ton block slipped and landed on his knees, and during a bad September storm a young seaman was thrown overboard and drowned. As autumn approached and the long daylight hours of summer receded, the artificers had to transfer to the rock and work by torchlight, adding to the general feeling of insecurity. Great was the relief when the season ended. Over the following winter the *Smeaton* made several trips to the granite quarries near Aberdeen and Peterhead, the masons in the Arbroath yard continued cutting and shaping stones for the 1809 season, and when weather allowed a small party visited the rock at low water springs to check that no disasters had occurred.

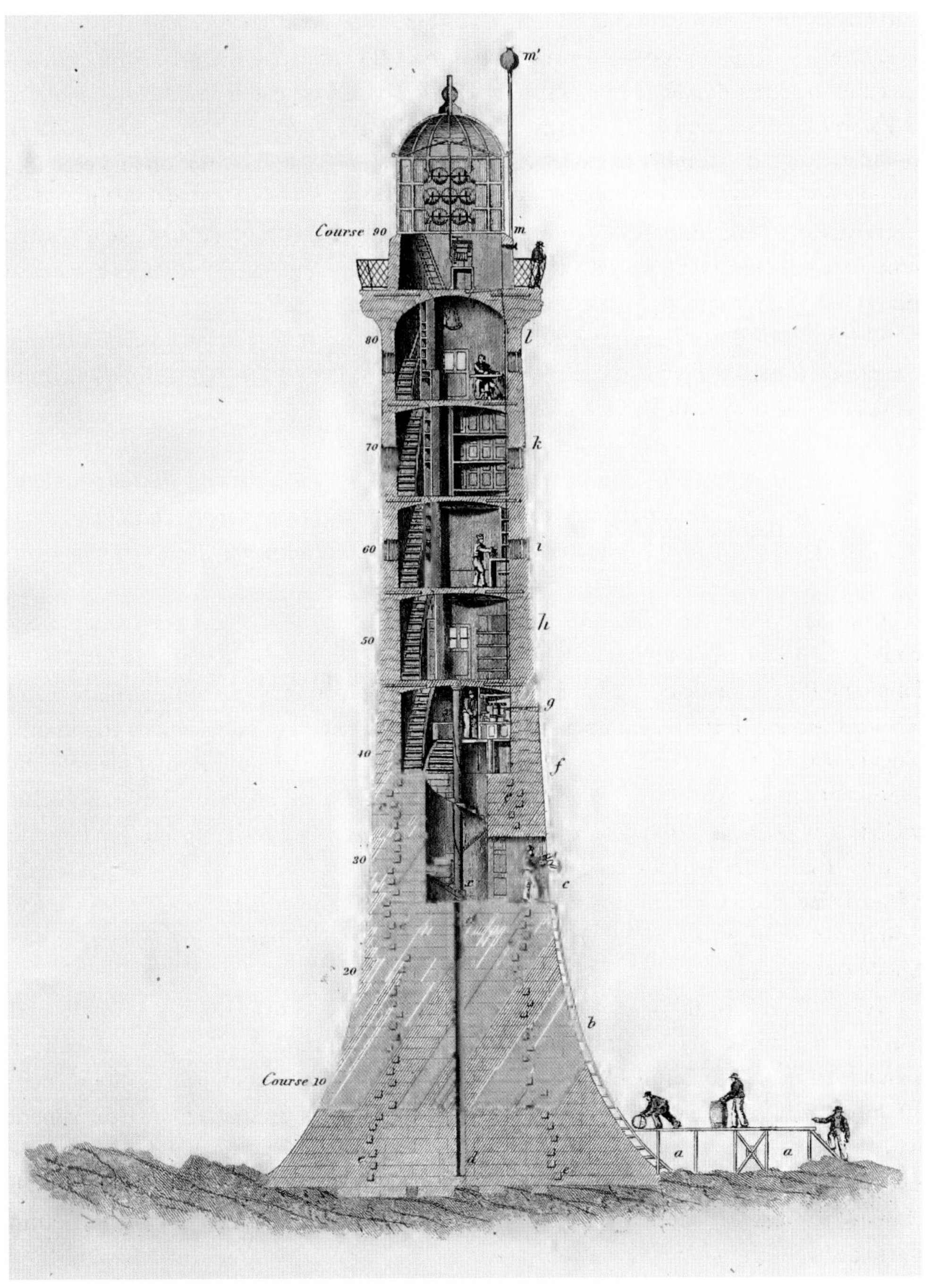

The master plan showing the lighthouse in its foundation pit.

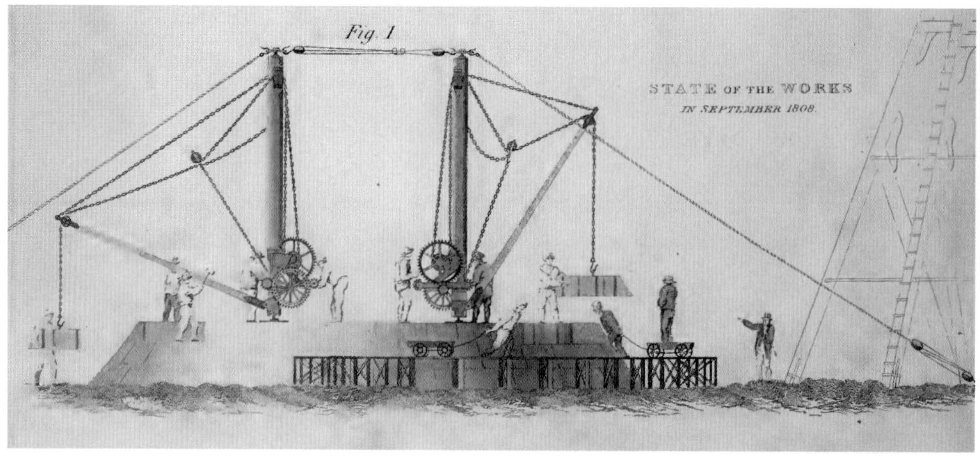

The state of the works in September 1808, showing the purpose-built cranes and three full courses nearing completion. The ladder and a timber leg of the raised Beacon House may be seen on the right.

1809 began threateningly and continued with its fair share of challenges. Heavy January seas damaged the railway's iron supports, loosened bracing chains on the Beacon House, and washed three huge boulders onto the rock, threatening work in progress. On the last day of May a heavy snowfall clothed everything in white, and in June the *Smeaton* had to run for shelter from a severe gale, leaving 11 artificers stranded in the Beacon House without bedding or food – a situation rescued by Robert's agreement to deliver 'a tea-kettle of mulled port wine'. On other occasions heavy stones were shifted off their beds, huge waves petrified artificers sheltering in the uncompleted Beacon House, and materials and equipment were washed away. Again there was human tragedy; Michael Wishart, a principal builder, was severely injured by a collapsing crane, and a large falling stone ended the life of labourer William Walker. But in one area personal safety and comfort were greatly improved. The Beacon House, initially an uncomfortable refuge in moments of crisis, was furnished to a reasonable standard for upwards of 20 artificers. Robert got his own personal apartment at the top and was clearly impressed:

> It was small, but commodious, and was found particularly convenient in coarse and blowing weather, instead of being obliged to make a passage to the Tender in an open boat, at all times, both during the day and the night, which was often attended with much difficulty and danger.

1809 was also a season of great engineering progress. By May many new courses of stones had been completed in Arbroath, ready for shipment. An extra tender, the sloop *Patriot,* entered service. On the rock, heavy seaweed growth was cleared from the stub in preparation for the fourth course of stones, and the railway was extended to encircle the tower completely. The men set up a rope bridge, affectionately known as Jacob's ladder, between the Beacon House and tower, allowing them and their mortar buckets to reach the growing stub even when it was partly submerged. As the rhythm of construction gathered pace, course after course was laid, and a benchmark was set on 20 August, when the 22nd course was completed in a single day. As Robert noted:

> This, as a matter of course, produced three hearty cheers. At 12 noon, prayers were read for the first time on the Bell Rock: those present, counting thirty, were crowded into the upper apartment of the Beacon, where the writer took a central position, while two of the artificers joining hands supported the Bible.

A few days later the 26th course was raised into position and the solid part of the tower, with its granite outer casing and sandstone core, was complete. The building now stood 31 feet 6 inches high, the daily tyranny of the tides was over, and Robert was delighted:

> The Bell Rock Light-house may therefore now be considered as from 8 to 10 feet above the weight of the waves; and, although the sprays and heavy seas have often been observed, in the present state of the building, to rise to the height of 50 feet, and fall with a tremendous noise on the Beacon-house, yet such seas were not likely to make any impression on a mass of solid masonry, containing about 1400 tons; its form being at the same time circular, and diminishing in diameter from the base to the top.

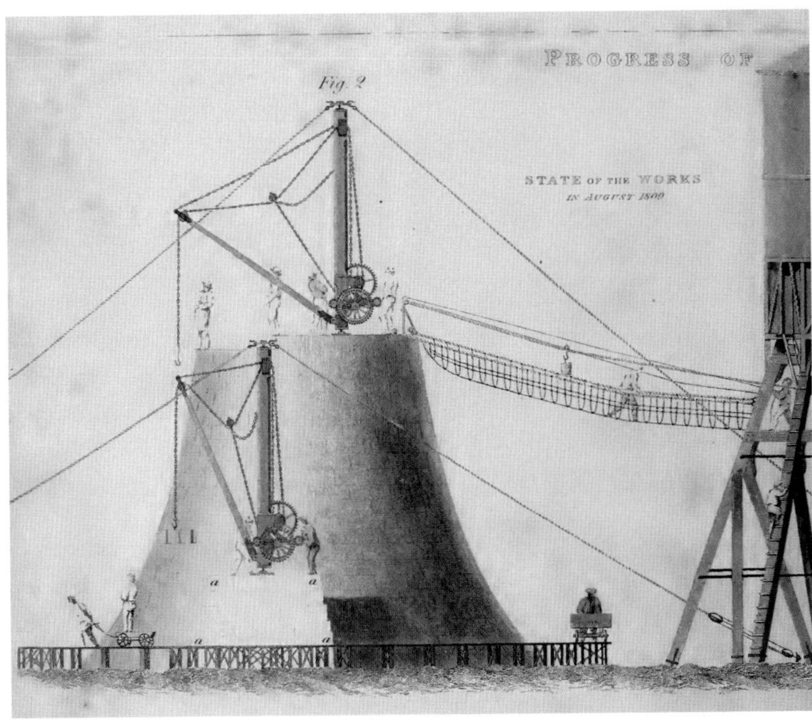

Nearing the end of the 1809 season: 'a mass of solid masonry, containing about 1400 tons'. Jacob's ladder connects the tower to the Beacon House.

It was a good moment to end building work for the season. From now on the tower would rise in sandstone only, its core hollowed out and fashioned to create store rooms and accommodation. Previous experience of September storms convinced Robert that he should wait until the spring before starting on the tower's entrance doorway and inner staircase – the 'void' as he called it. Just as important, the guyed crane used to hoist stones to the top had shown signs of instability and needed major redesign. As operations ended and the men returned to Arbroath, Robert was full of praise:

> In reflecting upon the state of matters at the Bell Rock, during the working months, when the writer was much with the artificers, nothing can equal the happy manner in which these excellent workmen spent their time. They always went from Arbroath to their arduous task cheering, and they generally returned in the same hearty state. While at the Rock, between the tides, they amused themselves in reading, fishing, music, playing cards, drafts, &. or in sporting with one another.

Robert returned to the Bell Rock in May 1810, intending to complete his lighthouse by the end of the season. The existing works had survived the winter storms well, stone courses up to the 44th were ready for shipment from Arbroath, a new crane balanced by a counterweight had been built and tested on land, and the previous year's successes gave him confidence that the final 700 tons of stone could be raised without major setbacks. But awful weather intervened. Unable to land on the rock and tossed around on a wild sea, Robert became increasingly worried about his schedule:

> These apprehensions were, no doubt, rather increased by the inconveniences of his situation afloat, as the Tender rolled and pitched excessively at times. This being also his first off-set for the season, every bone of his body felt sore, with preserving a sitting posture, while he endeavoured to pass away the time in reading; as for writing it was wholly impracticable.

At times like these doing gave way to thinking. Robert was finalising his design for the Light Room at the top of the tower, including the need to distinguish the light sequence from those of other Scottish lighthouses. He had been impressed by a previous visit to the Flamborough Head light in Yorkshire, which displayed alternating red and white beams, and decided on something similar. Orders had been issued to various manufacturers:

> The several compartments of the Light-room were now also in progress. The sheets of silver-plated copper for the reflectors having been ordered from Messrs Boulton and Watt,– the glass from the British Plate-Glass Company,– the cast-iron sash-frames from Mr John Patterson of the Edinburgh Foundery,– while the construction

> of the reflectors and reflecting-apparatus, together with the framing of the whole Light-room and its appurtenances, were executed under the immediate directions of Mr Thomas Smith, the writer's predecessor, who had now retired from the more active duties of engineer to the Light-house Board.

By late May the weather had started to cooperate and the Bell Rock became a hive of activity. Jacob's ladder was replaced by a sturdy wooden gangway. The much-improved balance crane entered service at the top of the tower, and the 27th course of 38 blocks, the first of the new season, was hoisted and laid with yet another three hearty cheers and a tot of rum. Work was now at the level of the tower's doorway and inner staircase, with walls 6 feet thick. The massive door lintel of 1.5 tons was soon ready for hoisting but faulty balancing of the crane caused some damage and a three-day delay. Yet by 5 June the 38th course had brought the building to a height of 45 feet, level with the first apartment, with walls now 3 feet thick. This was a key moment for the engineer, who had decided to stop using wooden trenails from this point upwards, saving time and money and making two courses per day possible. The men on the rock were now working a 9-hour day and earning princely wages of about £2 a week.

But as the tower grew higher and narrower the work became more dangerous. By the time they reached the 50th course the tower diameter had shrunk to 16 feet, the building was largely hollow, and the balance crane had to be manoeuvred upwards step by step. Robert was crucially aware of the responsibility he was placing on his foremen, and the risk that a fall would mean almost certain death. Furthermore, damage or loss of a single stone would stop work completely until a replacement could be cut and delivered from Arbroath.

Danger may have been focused on the tower, but it was the seamen marking time aboard the support vessels who gave Robert his next big headache. A mutiny was brewing and Robert had no doubt that their comparative idleness was to blame:

> It is, however, a strange, though not an uncommon feature in the human character, that when people have least to complain of, they are most apt to become dissatisfied, as was now the case with the seamen employed in the Bell Rock service, about their rations of beer.

A daily ration of six pints per man may seem reasonable enough, but there were malcontents aboard the tender and its captain advised Robert that 'if those who now complained were even to be fed upon soft bread and turkeys, they would not think themselves right'. Robert, never a man to mince his words or shirk an issue, lost no time in dismissing the ringleaders. The last thing he needed as his tower reached a critical stage of construction was a disruptive crew on the boats infecting the spirited workforce on the rock.

As June approached July the weather caused its own disruptions. On one occasion huge waves assaulted the tower and 'sprays fell from the heights in the most wonderful cascades, and streamed down the walls of the building in froth as white as snow'. On another, heavy seas washed over the top, gushing down through the interior and out through the entrance door. But building proceeded whenever Neptune permitted, and as course after course was added it became necessary to raise the stones in two stages using a winch and beam, or 'needle', extending from the storeroom window. By the end of June the 65th course was in place, forming the floor of the bedroom 80 feet above the sea. After six testing weeks on the rock, much of it spent in his tiny apartment in the Beacon House, Robert returned to Arbroath in contemplative mood:

> he had spent several weeks in a kind of active retirement, making practical experiment of the fewness of the positive wants of man. His cabin measured not more than 4 feet 3 inches in breadth on the floor; and though, from the oblique direction of the beams of the Beacon, it widened towards the top, yet it did not admit of the full extension of his arms when he stood on the floor; while its length was little more than sufficient for suspending a cot-bed during the night, calculated for being triced up to the roof through the day, which left free room for the admission of occasional visitants.

The engineer, in overall charge of men, money, and materials for one of the most demanding constructional projects ever attempted, had to make do with on-site accommodation roughly half the size of a 21st-century Scottish prison cell.

Back in the wide open spaces of Arbroath he checked that the workyard programme was nearing completion and decided it was time to pay off the stone cutters. Five guineas ensured a very lively party in the men's barrack, where they 'collected their sweethearts and friends, and concluded their labours with a dance'.

July 1810: the upper stonework nears completion.

Back on the rock another winch was installed at bedroom level, and on 9 July the last cargo of stones was delivered by the *Patriot* from Arbroath. As the building rose above the 80th course, its awkwardness increased. The tower began to flare outwards to accommodate the roof of the upper library and provide a footing for the Light Room. Robert decided to insert a circular flat-bar of Swedish iron, 3 inches deep by 1 inch wide, into the stonework, stabilising the building as it took the weight of the cornice and projecting stones. The 85th course, with stones over 7 feet long weighing more than a ton, proved especially cumbersome. But on July 30th the final piece of the jigsaw, the lintel of the Light Room door, was laid by Robert with a typical flourish. His tower now stood 102 feet high and, together with the Beacon House, timber bridge, winches, and railways, astonished the many visitors who ventured out from local ports in open boats to view one of the new century's most dazzling spectacles.

A dazzling spectacle: The Bell Rock works in July 1810.

No sooner was the masonry completed than the balance crane and two winches were taken down. On 4 August the weary but triumphant team of artificers – 18 masons, 2 joiners, 1 millwright, 1smith, 1 mortar-maker, and their foremen – prepared to leave the rock for Arbroath. Robert, proud and relieved in equal measure, paid his principal foremen generous compliments:

> All hands being collected and just ready to embark, as the water had nearly overflowed the Rock, the writer, in taking leave, after alluding to

the harmony which had ever marked the conduct of those employed on the Bell Rock, took occasion to compliment the great zeal, attention and abilities of Mr Peter Logan and Mr Francis Watt, foremen, Captain James Wilson, landing master, and Captain David Taylor, commander of the Tender, who, in their several departments, had so faithfully discharged the duties assigned to them, often under circumstances the most difficult and trying. The health of these gentlemen was drunk with much warmth of feeling by the artificers and seamen, who severally expressed the satisfaction they had experienced in acting under them; after which, the whole party left the Rock.

Later that day they were welcomed back in Arbroath by an enthusiastic crowd and Robert entertained his foremen, captains, and principal shore-based staff to a celebratory evening in the local inn, which ended by toasting 'Stability to the Bell Rock Light-house'.

There was still work to be done on the rock. The wooden bridge was taken down and the lighthouse staircase completed. Two cranes were installed to raise the cast-iron sash frames of the Light Room windows and when Robert returned from his annual tour of the Northern Lights on October 14th he found that excellent progress was being made. The Light Room was fitted with a copper cupola and 48 panes of window glass, completing the external works. It was just as well, for the weather was becoming boisterous, the nights were lengthening, and accommodation on the rock was increasingly 'cheerless'. A young artificer slipped, fell, and drowned. By November heavy storms and seas were assaulting the new lighthouse and on one occasion spray reached windows 104 feet above the rock.

But the Light Room was now waterproof and ready to receive the lighting system, supplemented by two external bells to warn mariners in foggy weather. The delicate light apparatus consisted of 24 oil lamps with parabolic reflectors, arranged on a rectangular frame that revolved once every 8 minutes. The two major sides of the frame had seven reflectors each, arranged in rows of two, three and two. The two minor sides had five each, arranged in rows of two, one and two. Panes of red glass placed in front of the minor-side reflectors produced the red flashes in the Bell Rock's unique light sequence.

It was only the second revolving light in Scotland and Robert insisted on the highest quality of materials and workmanship. Thomas Smith, now approaching 60, oversaw the work back in Edinburgh. Each parabolic reflector was fashioned from a copper sheet by 'a very nice process of hammering' and coated with silver. The illuminants were Argand lamps, patented by a Swiss inventor in 1780, with a circular wick housed in a glass chimney, gravity-fed with spermaceti (sperm whale oil) from a reservoir sufficient for 18 hours' continuous burning. The winding mechanism, which also struck the hammers of the fog bells, was powered by a drum and a weighted rope which descended right down through the tower.

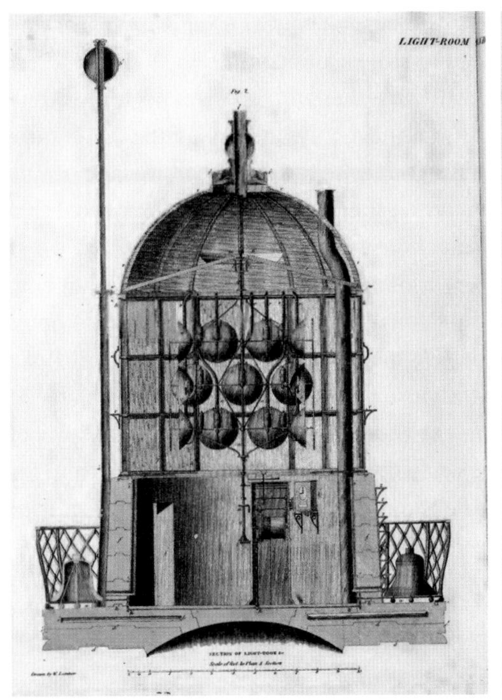

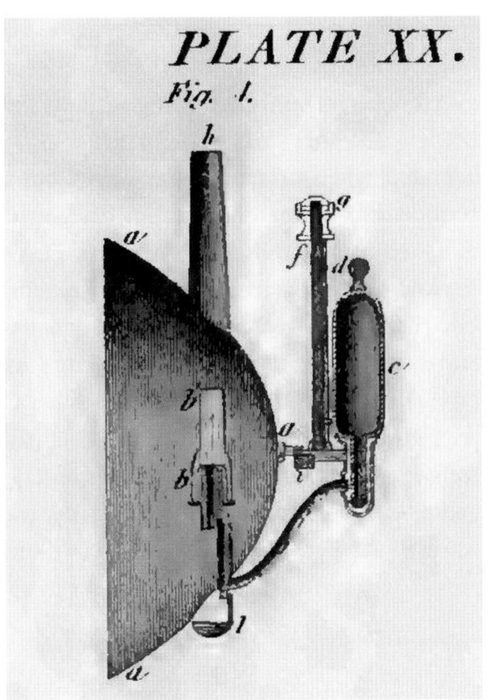

The Light Room with its revolving frame, lamps, winding mechanism, and fog bells. On the right is shown one of the Argand lamps with its parabolic reflector and oil reservoir.

By mid-December 1810 Robert was confident that the complete system would be ready within weeks and a historic public notice was issued:

> The Commissioners of the Northern Light-houses hereby give notice, That the Light will be from oil, with reflectors, placed at the height of about 108 feet above the medium level of the sea. The light will be exhibited on the night of Friday, the 1st day of February 1811, and each night thereafter, from the going away of day-light in the evening until the return of daylight in the morning. To distinguish this light from others on the coast, it is made to revolve horizontally, and to exhibit a bright light of natural appearance, and a red-coloured light, alternately, both respectively attaining their greatest strength or most luminous effect in the space of every four minutes … . During the continuance of foggy weather, and showers of snow, a Bell will be tolled, by machinery, night and day, at intervals of half a minute.

By the end of the year the lighthouse was in possession of its newly-appointed keepers and on 1 February 1811 the light shone forth as promised, settling the unequal contest between seafarers and the vicious reef that had plagued Scotland's east coast for centuries.

There was still plenty to be done: removing the beacon tower; establishing working routines for the keepers and the timely provision of stores; and completing a signal tower in Arbroath for communicating with the lighthouse together with apartments for off-duty keepers and their families. On the rock there was inevitably anxiety as the lighthouse 'vibrated' when struck by huge waves, but as time went by the keepers grew increasingly confident that Robert Stevenson's design would not let them down and even reported that their accommodation was as dry and warm 'as any house in Edinburgh'.

And so the construction phase of one of the world's iconic lighthouses came to an end. The challenges – technical, organisational, and personal – faced and overcome by Robert Stevenson almost defy our modern imagination. He had dreamt of solving the Bell Rock problem for many years before the Northern Lighthouse Board put John Rennie in charge of the project, and then proceeded to sideline his famous competitor with a dazzling display of technical competence, determination, and devotion to duty. Perhaps we may forgive him for ending the dedication of his 1824 *Account of the Bell Rock Lighthouse* to King George IV with words which by today's standards seem unnecessarily obsequious and self-congratulatory:

> It cannot fail to be gratifying to Your Majesty to learn, as the result of the exertions of this Board, that the mariner may now navigate those regions with a degree of security and confidence quite unknown to Your Majesty's Royal Ancestor JAMES THE FIFTH, when he sailed around this coast in the 16th century, or even, at a recent period, to Your Majesty's Royal Brother WILLIAM HENRY Duke of Clarence, when in early life he traversed those seas.
>
> With unfeigned sentiments of loyalty and attachment, the author subscribes himself,
>
> Your MAJESTY'S
>
> Most devoted Subject and Servant,
>
> ROBERT STEVENSON.

Others expressed themselves rather differently. J.M.W. Turner's famous etching of the lighthouse in a storm appears as a frontispiece to Robert's *Account;* and among the many plates is a sensitive, though far less well known, drawing by Robert's only daughter Jane, who acted as his secretary, and a poem entitled *Pharos Loquitur* ('The lighthouse speaks') by Sir Walter Scott, penned during a visit to the rock in 1814.

And what of young Alan, Robert's son, who was just 4 years old when the Bell Rock was first lit and 17 by the time his father's *Account* was published? Basically, he grew up with it. Delicate as a child and far from robust as an adult, he seems an unlikely candidate to follow a formidable and single-minded father who spared himself, and others, no hardship. Yet Alan somehow went on to tackle the notorious Skerryvore, no doubt finding the Bell Rock enlightening and daunting in equal measure as he confronted the Hebridean storms.

A C C O U N T

OF THE

BELL ROCK LIGHT HOUSE

Drawn by Miss Stevenson. Engraved by J.Horsburgh.

Pharos loquitur

Far in the bosom of the deep
O'er these wild shelves my watch I keep
A ruddy gem of changeful light
Bound on the dusky brow of Night
The seaman bids my lustre hail
And scorns to strike his timorous sail

See page 530.

Jane Stevenson's drawing and Sir Walter Scott's poem.

The Bell Rock Lighthouse as it stands today (photo: Ian Cowe).

Part 2: Skerryvore

To the Isle of Tiree

We were waiting on the quayside at Oban for a CalMac ferry to the Isle of Mull, a mere 40-minute affair. But a local warned us just in time that our queue was headed for Barra in the Outer Hebrides via the small isles of Coll and Tiree. It could have been awkward – three hours to Coll, another hour to Tiree and, had we braved wild seas to Barra, ten more before we made it back to Oban. Hebridean hops are sometimes longer than intended.

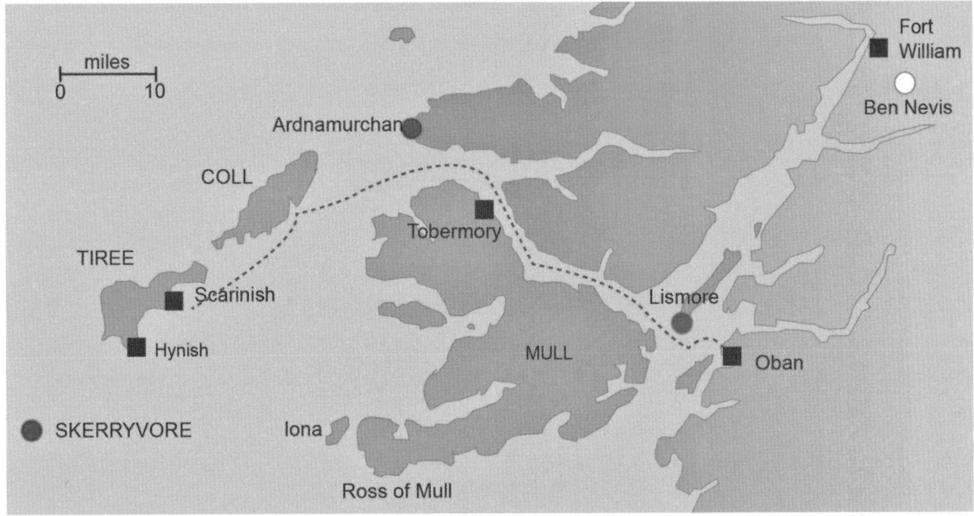

The ferry route from Oban to Tiree, showing the Lismore, Ardnamurchan, and Skerryvore lighthouses.

If, wisely, you catch a ferry from Oban to Tiree because you intend to, there's a treat in store. Robert Stevenson's 1833 Lismore lighthouse, the last he built as Engineer to the Northern Lighthouse Board, comes into view as you leave Oban, enticing the eye up magical Loch Linnhe towards Fort William and Ben Nevis. The boat squeezes through the Sound of Mull, hemmed in by highland drama, before leaving Tobermory harbour to port and heading out to sea. Look back and you will get a glimpse of Alan Stevenson's second most famous creation, Ardnamurchan, the westernmost lighthouse on the British mainland, embellished in the Egyptian style and lit in 1849, five years after Skerryvore.

After a brief call at Coll the ferry sails on to Tiree. Passing a glistening arc of sand at Gott Bay, it nudges up to the pier head close to the island's main village, Scarinish, a scattered collection of houses plus a food store, butcher's shop, and bank. Tiree is surprisingly flat for a Hebridean island, with its highest hill, Ben Hynish, just 463 feet above the surrounding ocean. Locals know their home as *Tir-fo-Thuinn*, the land beneath the waves.

Alan Stevenson's lighthouse at Ardnamurchan. The Isle of Rum is in the background (Wikipedia).

MV Lord of the Isles approaches the Isle of Tiree (photo: Tiree Images).

Another surprise is that Tiree, well clear of the mountains that encourage wet days on the mainland, enjoys one of the sunniest climates in Britain. It is treeless and exceptionally windswept, with spectacular silver beaches washed by an azure sea. Most of the island is covered by wind-blown shell sand, creating a well-drained machair with a stunning carpet of wildflowers in May and June. While the crofting community grazes cattle and sheep, the island is busy attracting solitude-seekers, windsurfers, birdwatchers, lighthouse enthusiasts, and fans enjoying a Skerryvore gig at the annual music festival.

Silver sands of Tiree, with the mountains of Mull in the distance (photo: Tiree Images).

Balephuil Bay, Isle of Tiree (Wikipedia).

But Tiree's present enchantment belies a turbulent history. Back in the second half of the 18th century the islanders were relatively prosperous and well fed (the Gaelic *Tiriodh* means 'land of corn'). Dr Samuel Johnson, giant of English letters and witty conversation, made a famous excursion to Scotland in 1773 at the age of 64, accompanied by his biographer James Boswell. Bravely subjecting his obesity and failing eyesight to a gruelling itinerary, some of it on a donkey, Johnson sailed to the islands of Mull, Coll, and Skye. Although he was unable to land on Tiree he learned of it while visiting neighbouring Coll and was clearly impressed by its ability to support a population of nearly 2000 souls:

> Near to Col is another Island called Tireye, eminent for its fertility. Though it has but half the extent of Rum, it is so well peopled, that there have appeared, not long ago, nine hundred and fourteen at a funeral.

It is unclear who counted the mourners, but he was right about the Isle of Rum. Forty miles north of Tiree, Rum has the classic Hebridean ingredients of high mountains, high rainfall, peaty moorland and red deer. Today its human population numbers less than 50 compared with Tiree's 650.

As is well known, the highlanders of Scotland suffered tragically during the 19th-century Clearances when large landowners, realising that sheep brought them more income and less trouble than people, forced their tenants off the land. Initially many were moved to the coast where they took up fishing and kelping – the production of industrial alkali from seaweed. But for most this was only a temporary reprieve and the economic woes caused by a sudden collapse of the kelp industry in the 1820's, together with ever-increasing population pressure, meant that living standards went into freefall. In fairness to the landowners, they were hardly to blame for the burgeoning population. Put bluntly, the fertility of the land could not keep up with that of its inhabitants. Many were forced to emigrate.

Less well known is that the Clearances affected the Hebridean islands just as much as the mainland. By 1804, when Robert Stevenson made his first inspection of the Skerryvore reef, the storm clouds were gathering over Tiree; its population had grown to 3,000 and in years of poor harvest the people went hungry. The situation was aggravated by primitive farming methods, scarcity of fuel, and absence of wheeled vehicles. By 1841, when Alan Stevenson was busy erecting his lighthouse tower on Skerryvore, the population was approaching 5,000 and the land was stretched to its productive limit. To add insult to injury the summer of 1846 brought potato blight, which had devastated the Irish potato crop the year before, and reduced Tiree to famine status. By the time Alan's *Account* of his lighthouse was published in 1848, the population had fallen to 3,200, 'assisted' by the Duke of Argyll and his notorious factor. A great exodus, much of it to Canada, was under way and continued for decades. Tiree's misery was profound.

It is hardly surprising that Alan, brought up and educated in Edinburgh and focused single-mindedly on the huge challenges of lighthouse construction, found the island uninspiring. It was the nearest land to the Skerryvore reef and he had no alternative but to base his operations there, but it totally lacked the skilled workers and facilities he needed. There were other disadvantages:

> This island is unhappily destitute of any shelter for shipping …. Nor is its interior more attractive; for although some parts of the

> soil when cultivated are excellent, the greater part of its surface is composed of sand. It was therefore obvious, at a glance, that Tyree was one of those places to which everything must be brought; and this is not much to be wondered at, as the population, who, on a surface not exceeding 27 square miles, amounted in 1841 to 4687 souls, labour under all the disadvantages of remoteness from markets, inaccessible shores and stormy seas, and the oft-recurring toil of seeking fuel (of which Tyree itself is destitute) from the Island of Mull, nearly 30 miles distant, through a stormy sea. It is said that this total absence of fuel in Tyree is the result of the reckless manner in which it was wasted, in former days, in the preparation of whisky; but, however this may be, certain it is that the want of fuel greatly depresses the condition of the people.

It seems a little out of character that Alan failed to mention at least one other 'factor' that was depressing the condition of the people during his years on Tiree.

And so, fast forward to happier times. Take the single-track road from Scarinish towards the south-eastern corner of the island and you come to Hynish, the centre of Alan's shore-based operations. It was here he decided to construct a pier, dock, and signal tower for communication with his lighthouse. The heavy ocean swells that pound the southern end of Tiree made it far from ideal, but at least the pier and dock entrance are reasonably sheltered by the south-eastern tip of the island, and the signal tower has an uninterrupted view across 12 miles of restless ocean to the spikes of Skerryvore.

The dock is still there today, undeniably impressive. Its unique feature is a freshwater flushing system fed by a reservoir and conduit to keep it clear of sand and debris. The entrance, 20 feet wide, was designed to take timber stop-booms lifted into position by a hand-operated crane, giving a safe mooring and shelter to the Skerryvore tender. Nowadays dock and pier are occasional hosts to local fishing and pleasure craft.

Alan Stevenson's dock and pier at Hynish (sketched from a photo supplied by Tiree Images).

The Hynish signal tower (sketched from a photo supplied by Tiree Images).

The signal tower was used to transmit and receive visual semaphore signals to and from the rock at pre-arranged times, or otherwise in an emergency. It remained in service until 1892 when the Commissioners decided to transfer the service to Erraid on the Isle of Mull. With the advent of reliable radio communication, the shore station was relocated to Oban on the mainland. The lighthouse was fully automated in 1994.

In 1984 the Northern Lighthouse Board offered the Hynish signal tower to a local charity, the Hebridean Trust, for conversion into a Skerryvore Lighthouse Museum. The Lighthouse Keepers' Cottages in the Upper Square, near the signal tower, were purchased in 1997 and restored to provide homes for four highland families. Buildings in the Lower Square, near the dock, were offered to the trust by the Argyll Estate in the late 1980s and have been restored to provide the Hynish Centre with visitor accommodation and group facilities for universities, businesses, and local functions. One of the buildings has been renamed Alan Stevenson House.

At the time of writing, I am about to experience Tiree and Skerryvore myself, armed with a wonderful book describing 'the most graceful lighthouse in the world', and I hope to take you with me. Alan's *Account* promises a fascinating insight into the mindset, technical knowledge, and courage of a Victorian engineer as he tackled his most formidable challenge.

Great preparations

The preface to Alan Stevenson's *Account* has none of the obsequious tones of his father's Bell Rock dedication to King George IV in 1824 – no mention of the sovereign (now Queen Victoria), no hint of self-congratulation for his immense technical achievement, just an apology for the book's shortcomings, an acknowledgement of the debt he owes his father, and fulsome thanks to his God for the preservation of life and limb:

> Although, in the course of the Narrative, I have occasionally noticed some special deliverances from danger, I have altogether neglected to record the remarkable fact, that, amidst our almost daily perils, during six seasons on the Skerryvore Rock, there was no loss of either life or limb amongst us. Those who best know the nature of the service in which we were engaged, – the daily jeopardy connected with landing weighty materials in a heavy surf and transporting the workmen in boats through a boisterous sea, the risks to so many men, involved in mining the foundations of the Tower in a space so limited, and above all, the destruction, in a single night by the violence of the waves, of our temporary barrack on the Rock, which had cost the toils of a whole season, will not wonder that I am anxious to express, what I know to have been a general feeling amongst those engaged in the work – that of heartfelt thankfulness to Almighty God for merciful preservation in danger, and for the final success which terminated our arduous and protracted labours.
>
> Edinburgh, March 25, 1848.

Yet Skerryvore is a masterpiece of Scottish engineering which surely equals, and in some respects exceeds, Robert's triumph in the North Sea. It is clear that in Alan Stevenson we are dealing with a far more complex and vulnerable character.

Alan's father had first landed on the Skerryvore reef in 1804 during one of his annual voyages around the coast as Engineer to the Northern Lighthouse Board. He inspected it again in 1814, accompanied by Sir Walter Scott and a party of Commissioners. In the same year Parliament sanctioned the building of a lighthouse there, but the Board was in no great hurry. Already committed to a major building programme of lighthouses in more accessible places, the Commissioners were hampered, as always, by a shortage of funds. In spite of the Bell Rock's success and growing fame, an even more dangerous and expensive adventure 12 miles off the coast of Tiree was not to their taste. Another 20 years were to pass before the pressure to act became irresistible and then, in 1834, they decided to send Robert back to Skerryvore. This time two of his sons, Alan and Tom, went with him.

The fronticepiece to Alan Stevenson's book.

ACCOUNT

OF THE

SKERRYVORE LIGHTHOUSE,

WITH

NOTES ON THE ILLUMINATION OF LIGHTHOUSES;

BY

ALAN STEVENSON, LL.B., F.R.S.E., M.I.C.E.,

ENGINEER TO THE NORTHERN LIGHTHOUSE BOARD.

"ΥΠΕΡ · ΤΩΝ · ΠΛΩΙΖΟΜΕΝΩΝ"

Inscription on the Ancient Pharos of Alexandria.

BY ORDER OF THE COMMISSIONERS OF NORTHERN LIGHTHOUSES.

ADAM AND CHARLES BLACK, NORTH BRIDGE, EDINBURGH.
LONGMAN AND CO., LONDON.

MDCCCXLVIII.

Robert's instructions were to make a detailed survey of the whole seven-mile reef and investigate the southern coast of Tiree to establish the best location for a signal tower and dock. His report to the Commissioners confirmed the Skerryvore rock outcrop as the only suitable site for a lighthouse tower even though its surface was highly irregular and deeply undermined by subsea fissures. He also recommended Hynish for the shore base. Never one to duck a challenge, Robert concluded that the lighthouse was not only practicable but promised to be 'much less difficult and expensive than that of the Bell Rock'. This surprising claim, which glossed over the total lack of facilities on Tiree and the special hazards of Hebridean storms, probably had more to do with his impatience for action than a realistic appraisal of the technical and organisational difficulties. In any case the Commissioners decided to visit Tiree and Skerryvore themselves in the summer of 1835, and it was perhaps providential that a fire broke out in their steamer's boiler room close to the dreaded rocks, helping clinch the argument for a lighthouse. But the person they put in charge of the project was not the ageing Robert; it was his 28-year-old son.

So what had Alan, the pale and sickly child who had given his mother such anxiety, been doing in the meantime to justify this resounding vote of confidence? Basically he had been preparing, with considerable misgivings, for life as a professional lighthouse engineer. Or perhaps it would be fairer to say that as the eldest son he was being 'prepared' by his father to further the interests of the family's Edinburgh business and the Northern Lighthouse Board. But this particular child was destined to become a man of many parts. As adulthood approached he felt more and more troubled by the thought of committing himself wholeheartedly to a profession demanding relentless mental and physical effort.

Alan grew up in the large and fashionable home, 1 Baxter's Place in Edinburgh's New Town, that Robert had built for his family as his fame and income expanded. It was a wonderful place for children, with masses of space and a large garden in which Alan and his younger brother Tom had a lot of mischievous fun. But it was ruled over by a father with an old-fashioned Christian outlook, a superhuman work ethic, and strong views about the education of his sons. In 1815, at the age of eight, Alan was dispatched to the local Royal High School and fed the usual diet of classics and natural philosophy. Rather shy and introspective, he failed to shine in spite of his obvious intelligence, preferring literature and poetry to physical activity and giving his father increasing concern. When his education continued at Edinburgh University he remained torn between the arts and sciences, but Robert kept up the pressure and eventually received the news he wanted to hear. Alan agreed to join the family business.

The decision set in motion a programme that mirrored Robert's own introduction to practical engineering with his mentor, Thomas Smith, more than 30 years before. Alan's seven-year apprenticeship, even broader than his father's, was buttressed by scientific knowledge that eclipsed Robert's more tentative efforts at formal education. The business was now handling a wide range of engineering projects including roads, bridges, harbours, and railways, so Alan spent summers on projects scattered far from their Edinburgh base and winters helping his father with a diverse and ever-increasing office workload. He was heavily involved with Robert's lighthouses at the Rinns of Islay (lit in 1825), Buchan Ness

(1827), Cape Wrath (1828), Tarbat Ness (1830), and Mull of Galloway (1830); and found himself, at various times, dispatched to bridge works near Dumfries, harbour works in Fife, and river works near Liverpool under the supervision of the famous Thomas Telford. He also travelled to France, Russia, and Sweden, mixing engineering with sightseeing and a fascination with foreign lands. And then in 1830 the Northern Lighthouse Commissioners, realising that they had an exceptional young man in their sights, offered him a job as their Clerk of Works at a salary of £150 per annum. He was just 23 years old.

Alan's workload, already heavy, now increased dramatically. Five new lighthouses including Robert's final gem, Lismore (1833), were built and lit in the next few years, and it was the job of the Clerk of Works to oversee the ordering and delivery of materials, ensure the quality of workmanship, concern himself with safety, and generally check that the works were being carried out in the Board's best interests. In view of his history of indifferent health it seems remarkable that he survived such a severe crash course in responsibility for men, money, and materials.

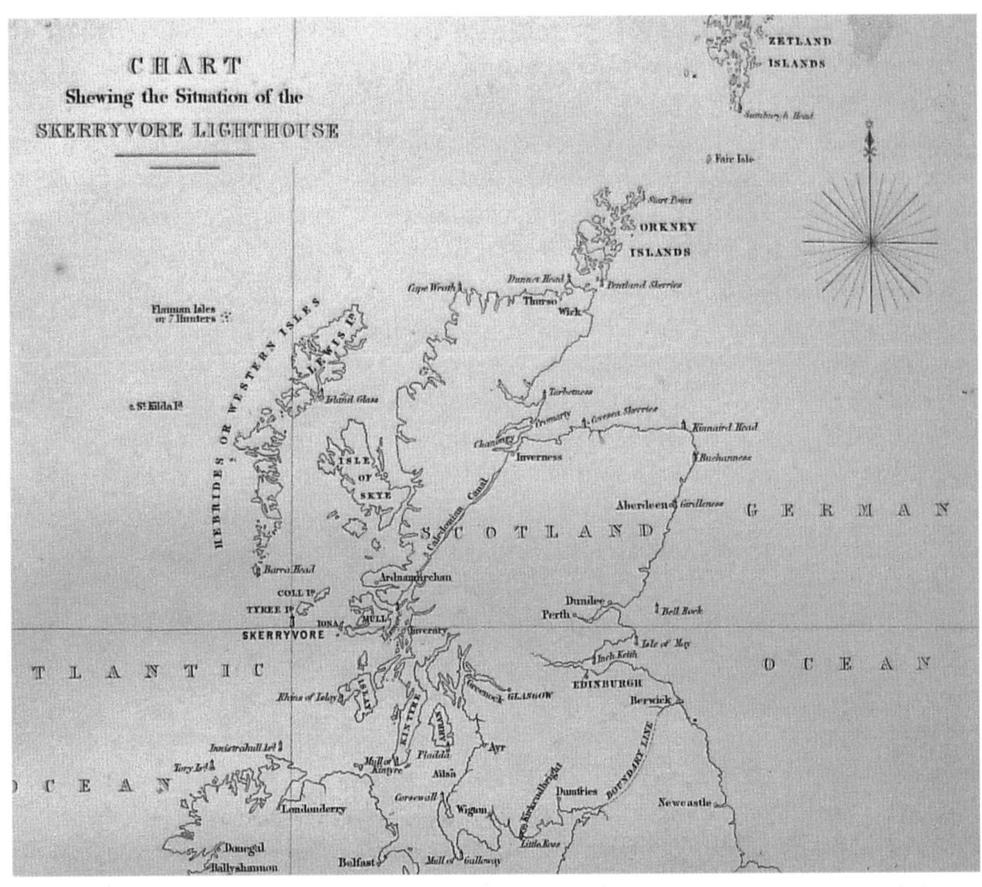

Alan Stevenson's 1848 chart showing the location of Skerryvore and the neighbouring lighthouses at Rhinns of Islay and Barra Head. Note also the Bell Rock in the German Ocean (North Sea).

All this is what Alan had been doing, under the eagle eye of his father, prior to Skerryvore. The programme of education and practical experience gave him a knowledge of lighthouse engineering unsurpassed by any other man of his age. Unsurprisingly, the Commissioners decided in 1838 to offer their Clerk of Works an additional six guineas a week to take charge of the Skerryvore lighthouse project. They would not regret their choice. And five years later, with Skerryvore nearly completed, they appointed him Engineer to the Northern Lighthouse Board.

The opening pages of Alan's *Account*, published four years after Skerryvore was lit, are devoted to scene-setting: the constitution of the Lighthouse Board; lights established since 1821; brief notes on systems of illumination, beacons and buoys. And then, in Chapter 1: 'Topographic notice of the Skerryvore Rock', he raises the curtain on Skerryvore by describing the vicious reef that disturbed his waking – and no doubt sleeping – hours over a period of six years. Its precise features had a huge influence on the whole project.

After recalling his father's visits to Skerryvore in 1804, 1814, and 1834, when the first detailed survey of the rocks was undertaken, Alan eloquently describes the historical background of lost ships and the practice of local fishermen to visit the rocks after gales 'in quest of wrecks and their produce, in finding which they were too often successful'. He then lets loose on the Skerryvore reef:

> a tract of foul ground, consisting of various small rocks, some always above the level of the sea, others covered at high water, and exposed only at low water, and others, again, constantly under the surface, but on which the sea is often seen to break after heavy gales from the westward. This cluster of rocks extends from Tyree in a south-westerly direction, leaving, however, between that island and the rock called Boinshly, the first of the great Skerryvore cluster, a passage of about five miles in breadth ...This passage is called the passage of Tyree; but it is by no means safe during strong and long continued gales, as the sea which rises between Tyree and Skerryvore, is such that no vessel can live in it. I have myself often seen it one field of white broken water, the whole way from Tyree to the Rock; and we know that the wreck of the Majestic, which occurred in 1841, during the progress of the works, was entirely caused by the heavy seas which she encountered off Boinshly.

The passage of Tiree, in which 'no vessel can live', is precisely the stretch of water that Alan and his men would negotiate time and time again as they struggled to Skerryvore and back.

Three miles from Skerryvore (derived from the Gaelic *Sgeir Mhor* or 'Great Rock') lurks treacherous Boinshly ('Deceitful Bottom'), which was quite capable of drawing a boat onto it 'by a kind of suction' and shooting columns of water upward like a gigantic fountain. It was not Skerryvore's only neighbour. In between stood Bo-rhua ('Red Rock'), covered at high water but dry at low, with a nasty little outlying pinnacle about ten feet square. Together they formed a formidable boat-splintering trio. And three miles seaward lay more rocky traps for the unwary, to be named Mackenzie's, Fresnel's, and Stevenson's, often under water but betrayed, if at all, by swirling seas above.

Skerryvore itself was the only rock cluster offering a viable platform for a lighthouse. Fortunately it stood about 12 feet above high water even at spring tides, meaning that workmen would not have to vacate it every time the tide rose. The largest surface exposed at low tide was about 280 feet square, but it was far from level and the usable area was considerably reduced by deep gullies or fissures, one of which undermined the rock and terminated in a hollow submarine chamber capable of ejecting a beautiful but threatening 20-foot water spout.

The rocks interested Alan not only for their engineering qualities – extremely hard gneiss containing quartz, feldspar, hornblende, and mica – but for the few sea plants that somehow managed to cling on, the storm-worn pebbles and reddish-brown centipedes in the crevices, the hard guano-like substance found in rock pools around high water mark, the seabirds and seals. He was always sensitive to the natural world in all its beauty and grandeur.

The action of the sea had worn Skerryvore glassy-smooth, more like a mass of dark splintered glass than a rocky reef. Landings in rough weather were extremely risky. Initially there were many accidents, fortunately none fatal, and the foreman mason was heard graphically describing landing on the rock as 'like climbing up the side of a bottle'. There was general relief when Alan had the widest and deepest gully cleared and dressed by blasting under water to form a proper landing creek.

It was not only the rock itself that mattered. The rise and fall of the tides at Skerryvore and Hynish, their height as well as their timing, and the strength of ocean currents, were to form a never-ending backdrop to hundreds of arrivals and departures of men, materials, tools, provisions, and thousands of tons of dressed stone for the lighthouse tower. Alan estimated a tidal range of between 12 and 13 feet at high spring tides and about 3 feet at neaps, with maximum current velocities between 4 and 5 miles an hour. And even when all this had been taken into account, there were waves, winds, and an almost perpetual Atlantic swell to deal with – a scene starkly and sometimes brutally different from the flat calm portrayed in the frontispiece to his *Account*.

Great preparations were needed before lighthouse construction could begin. In Chapter II: 'Preliminary arrangements and works, including survey of the rocks, and opening of quarries, from 1834 to 1837', Alan covers two of the most important: a detailed survey of the whole Skerryvore reef to protect boats and their crews as they sailed to and from Hynish and manoeuvred close to the rocks, often in deteriorating weather; and a search for suitable stone for the tower given that Tiree was, in this respect as in many others, unlikely to give satisfaction.

The survey, begun in 1834 and completed in 1835, was far more detailed and accurate than anything attempted before in the Hebrides 'in regard to surface and levels, which are always necessary for the purposes of the Engineer'. Basically it involved extensive triangulation using shore landmarks and moored buoys placed at convenient points,

Skerryvore (photo: Ian Cowe).

to pinpoint more than 140 rocks in and around Skerryvore. Their locations were complemented by 500 depth soundings. The work was painstaking and risky but it yielded an enormous prize, with the:

> exact details of the depths, rocks, and shallows of the surrounding sea, with the nature of the bottom, accurately laid down; and our experience during the course of the work, more than once shewed how essential was the possession of minute topographic information to the safety of the shipping attending the works; more especially as some of the vessels lay very near the rocks, and were frequently driven, by a sudden change of wind, to seek shelter, during the darkest nights, among the neighbouring islands.

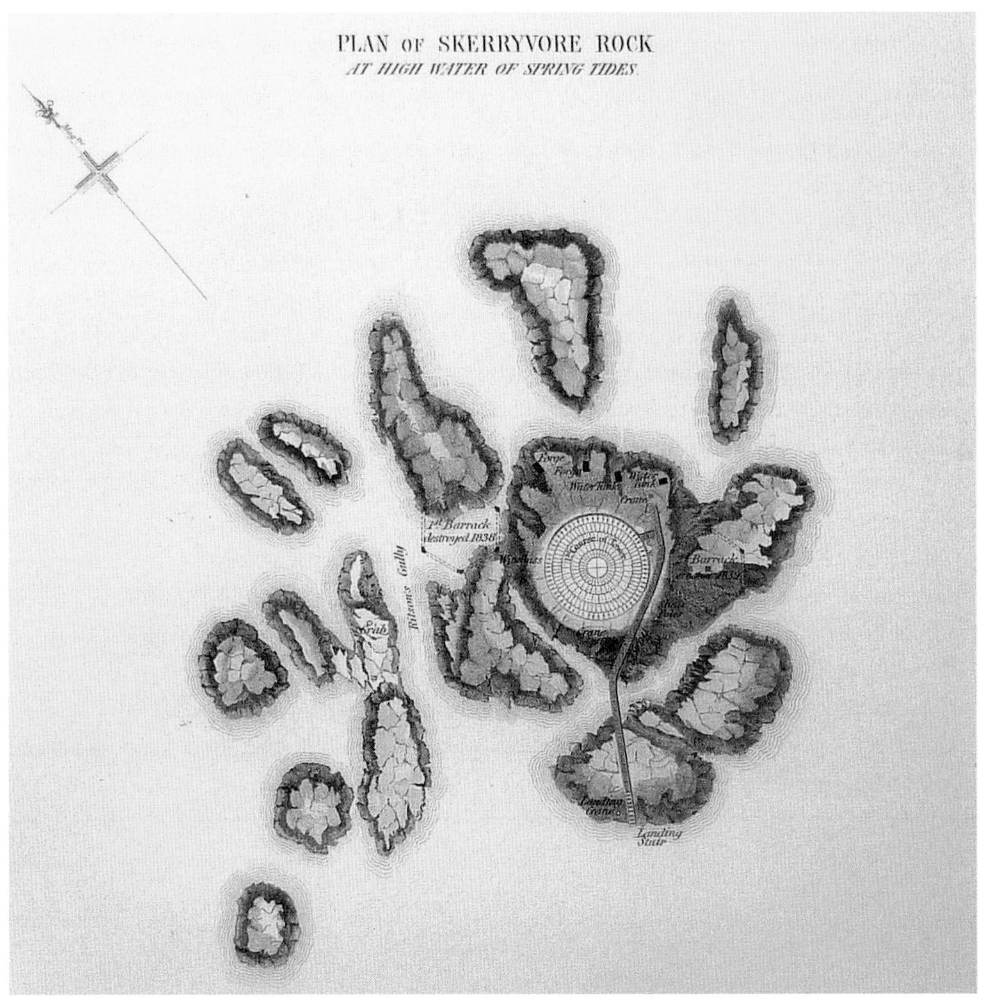

Fruits of the survey: a detailed plan of the Skerryvore cluster at high water of spring tides, showing the eventual location of the tower.

The other major problem was stone for the tower – where on earth would it come from? Robert, in his Bell Rock project 30 years before, had been able to call on stone from a number of well-established quarries on Scotland's east coast, but Alan's Hebridean sources were limited and their suitability unknown. Fortunately the Duke of Argyll agreed to grant the Commissioners permission to quarry materials on any part of his extensive estates and as a first step a party of 14 quarriers was sent to investigate the gneiss rocks close to Hynish Point:

> In the summer of 1837, Mr Scott and his party turned out about 3800 cubic feet of rock, capable of being applied to the purposes of squared masonry, and a very large quantity of stones fit for rubble work. This produce, although small, if contrasted with that of established quarries, is by no means despicable, when the force employed and all the disadvantages of the situation are considered; and if the nature of the material, which is full of rents and fissures (technically called dries and cutters), the frequent deceptions attending the opening of new quarries, the excessive hardness and unworkable nature of the rock, the quality and size of the blocks required to entitle them to claim a place in a marine tower, and the great loss of time, caused by the badness of the weather, be considered, it will not appear that Mr Scott and his party had been eating the bread of idleness.

Not to put too fine a point on it, quarrying stone at Hynish was a nightmare. The yield of 3,800 cubic feet was very meagre bearing in mind that the solid base of the tower alone, due to rise 26 feet above its foundation pit, would need about 27,000 cubic feet. It was clear to Alan that the gneiss of Tiree could hardly be expected to provide more than the first few courses, and that another source of stone would have to be found, preferably not too far away. Fortunately the search could be postponed until the following year – and it would lead him to the Isle of Mull.

The Duke of Argyll was already proving a firm supporter of a Skerryvore lighthouse. Struggling to cope with his Tiree tenants as their numbers rose and the food supplies dwindled, he probably welcomed the distraction. In any case he agreed to allocate a total of 45 acres at Hynish for a workyard, harbour, and buildings to accommodate the works and eventually house the families of the lighthouse keepers and the crew of a tender vessel. By the end of 1837 the momentum was unstoppable and the Commissioners formed a committee to superintend erection of the lighthouse, with Alan as their resident engineer.

‘On the Construction of Lighthouse Towers’

The title of Alan’s third chapter puts down a marker for the remainder of his *Account.* Unlike John Rennie and his father, who had agreed to base the Bell Rock firmly on Smeaton’s design for the Eddystone with its copious dovetails, trenails, and joggles, Alan was determined to return to first principles and ask what was actually necessary to ensure the stability of an isolated stone tower assaulted by storm-force waves.

Robert Stevenson was a courageous and practical man with a supreme confidence in his own engineering instincts. His account of the Bell Rock is packed with the dangers and excitements of lighthouse engineering on a wave-washed rock, but he spends little time explaining his or Rennie’s design decisions. Of course it is only fair to add, more than 200 years later, that the choices made seem entirely justified – even if they did owe a lot to Smeaton’s pioneering example.

The civil engineering profession was in its infancy in the late 1700s. The first formal meeting of British civil engineers took place at a London tavern in March 1771 with John Smeaton as its leading light. However in time his Society of Civil Engineers morphed into an exclusive dining club rather than a professional body and a younger group of men began to press for something more useful. The Institution of Civil Engineers was founded in 1818 and the great Thomas Telford agreed to become its first President in 1820, remaining in post for 14 years. His involvement was crucial because he was determined it should not become a mutual admiration society, nor an employment agency for engineering talent, but rather a fount of engineering knowledge and achievement that would promote civil engineering as a profession.

By the time Alan Stephenson came to tackle Skerryvore civil engineering was well established, with a fast-growing body of knowledge on design and construction in stone and iron. Rock lighthouses were hardly mainstream, but even here experience of the effects of water on stone structures including bridges, docks and harbour walls might prove useful. The modest self-effacing Alan, graduate of Edinburgh University, fellow of the Royal Society of Edinburgh and member of the Institution of Civil Engineers, was hardly one to launch semi-blindfold into a project long considered impossible without carefully reviewing and weighing the technical background.

In 25 fascinating pages Alan takes us through the thought processes that decided the size, shape, and structural details of his lighthouse. He divides the various considerations into two classes:

> 1st, Those which refer to elements common to Lighthouses in all situations, and differ only in amount, such as the height of the Tower necessary for commanding a given visible horizon, and the accommodation required for the Lightkeepers and the Stores; and

> 2nd, Those which are peculiar to Towers in exposed situations, and which refer solely to their fitness to resist the force of the waves which tend to destroy them.

Deferring a decision about tower height until later in the chapter, he notes that a lighthouse on an isolated rock, cut off from the land by storms for long periods, needs more internal space for men, provisions, and stores than a shore-based building – not least because the Board's policy was to have three or even four keepers on duty rather than the usual two:

> In the long nights of a Scotch winter, when the lamps are kept burning for about seventeen hours, during which time they are never left for a moment without the superintendence of at least one Keeper, the care of the light, even in the most favourable situations, necessarily occupies at least two persons; but in places like the Eddystone, the Bell Rock, and the Skerryvore, where it sometimes happens that six or eight weeks elapse without its being possible to effect a landing, it has been thought necessary that there should never be fewer than three Keepers on duty.

This was clearly a prudent safeguard in case one of the keepers became ill or was injured; and it has often been rumoured that in the case of a death two survivors would provide a more convincing assurance that no dastardly deed had taken place.

And now comes the real meat of the chapter: the 'second class of considerations', which refer to the survivability of the building – the focus of the engineer's dreams and nightmares, the overriding criterion by which he will be judged by colleagues and history, and the reason he has been entrusted by the Northern Lighthouse Board to confront the Hebridean storms. True to his nature, Alan starts with an admission that his father could never have made:

> The first observation which must occur to any one who considers the subject is, that we know little of the nature, amount and modifications of the forces, on the proper investigation of which the application of the principle which regulates the construction must be based. When it is recollected, that, so far from possessing any accurate information regarding the momentum of the waves, we have little more than conjecture to guide us, it will be obvious, that we are not in a situation to estimate the power or intensity of those shocks to which Sea Towers are subject; and much less can we pretend to deal with the variations of these forces which shoals and obstructing rocks produce, or to determine the power of the waves as destructive agents.

56

He tries to rescue the situation by citing the only available evidence on the subject, obtained by his brother 'Mr Thomas Stevenson, Civil-Engineer', who developed a 'marine dynamometer' for assessing the maximum pressure exerted by ocean waves. In experiments around Tiree and Skerryvore in 1843–4, Tom measured an average value of 611 pounds per square foot during the summer months, rising to 2086 in winter, with a peak value of 4335 pounds per square foot recorded at the Skerryvore rock. This figure has been widely quoted ever since – even though Tom went on to record 6,083 pounds per square foot the following year during a severe westerly gale. Either peak value is, of course, fearsome: a pressure of around 2 to 3 tons on every square foot exposed to the waves. But how accurate are the measurements and what exactly do they mean for the stability of a lighthouse tower?

Thomas Stevenson's marine dynamometer.

Tom's dynamometer consisted of a cylinder that was firmly bolted to the rock where measurements were to be made. The waves impinged on a circular plate (A), typically 6 inches in diameter, fitted with four guide rods (B) that entered the cylinder and were attached to a powerful steel spring. Wave forces on the plate pushed the guide rods into the cylinder, extending the spring, and the maximum extension was 'recorded' by four non-return sliding leather rings (T) mounted on the guide rods at the far end of the cylinder. At the conclusion of the experiment the positions of the rings were read off and averaged, and a simple calculation revealed the maximum pressure in pounds per square foot exerted on the circular plate. The device was generally placed at about three-quarters tide; any lower, and it would often have been impossible for anyone to approach it for days on end to investigate the reading. Tom conducted no less than 267 experiments around Tiree and Skerryvore, and on the Bell Rock, the latter yielding a maximum of 3,013 pounds per square foot, about half the peak he obtained on Skerryvore in 1845.

Tom's efforts were original and painstaking, but he himself admitted that:

> there may be some objection to referring the action of the sea to a statical value. Although the instrument might perhaps be made capable of giving a dynamical result, it was considered unnecessary, in these preliminary experiments, to do anything more than represent the maximum pressure registered by the spring.

The lack of a 'dynamical result' was surely a limitation. In a laboratory experiment the dynamometer, if loaded with a slowly increasing force up to some maximum value, could be expected to record it accurately; but mounted on a rock in the sea, pounded by pulsating waves, its dynamics would surely have come into play. A mass-spring system normally displays one or more natural frequencies at which it tends to vibrate, and movements caused by repetitive impulsive forces may well exceed those due to steady, or 'statical', ones. Who can tell what complicated patterns of spring extension were produced by Skerryvore's violent waves which, as Tom admitted, were 'more or less broken by hidden rocks or shoal ground before they reached the instrument'?

Another problem concerns 'edge effects'. Wave forces would not be generated evenly over the surface of the dynamometer's circular plate, especially near its edge as water spilled over. Yet an even distribution was assumed when calculating the pressure in pounds per square foot. The discrepancy would be worse for a small plate than a big one because the edge length (circumference) is proportional to the diameter whereas the area is proportional to the diameter squared. And any effects caused by water striking the large curved surface of a stone tower would presumably be quite different from those experienced by a small flat plate.

It is only fair to add that Alan and his brother were aware of the dynamometer's limitations, and Tom even went on to experiment with impulsive forces using the 19th-century expedient of dropping cannon balls on the instrument. But this was well after Skerryvore was built and it seems that Alan's use of Tom's results reflects a desire to confirm his tower design after the event, applying a quantitative veneer to decisions that had actually been based more on instinct and experience. Alan is honest enough to suggest this towards the end of his dynamometer discussion. After quoting the peak pressure of 4,335 pounds per square foot recorded by Tom on Skerryvore in 1843–4 (but not the 6,083 value obtained a year later), he adds:

> But these experiments have not been continued long enough as yet to render them available for the Engineer. In the present state of our information, therefore, we cannot be said to possess the elements

> of exact investigation, and must consequently be guided chiefly by the results of those numerous cases which observation collects, and which reason arranges, in the form which constitutes what is called professional experience.

The pool of experience needs expanding by prolonged observation of ocean waves and their effects:

> their magnitude and velocity, their level in regard to the rocks on which they break, the height of the spray caused by their collision against the shore, the masses of rock which they have been able to move, and those which have successfully resisted their assault … in any given case, the problem is to be solved chiefly by the union of an extensive knowledge of what the sea has done against man, and how, and to what extent, man has succeeded in controlling the sea … nor does it seem possible as yet to found the art of Engineering, in so far as it refers to this class of works, upon any more exact basis … time only can test the success of our schemes in cases of difficulty.

Careful measurements, the steady accumulation of knowledge about what worked in the past – ingredients that the Institution of Civil Engineers, founded 30 years before, might hopefully have supplied. But wave-washed lighthouse towers were still out on a limb, and measurement of ocean waves and their effects on man-made structures had barely moved in the 90 years since John Smeaton had risked his reputation on Eddystone and the 30 years since Robert Stevenson had crowned Bell Rock. Tom's experiments were performed when Skerryvore was almost completed and could at best only provide a retrospective and somewhat lukewarm endorsement. The harsh truth is that Alan's tower design was only supported by the practical certainty that Rennie and his father had not made any dreadful mistakes. All calculations were his own; and any departure from their way of doing things would require a lot of courage.

There is another topic – wave statistics – that Alan could not have known much about. Today ocean waves are extensively studied, with an impressive theoretical background. We know that wave heights, even in a nominally stable sea state, are highly variable and that occasional waves may exceed the average by a large margin. Dangerous 'rogue' waves, sometimes called hundred-year waves, occur randomly. Skerryvore, facing Newfoundland across 2,000 miles of wild Atlantic, could certainly expect occasional rogues, but Tom's dynamometer was unlikely to meet one during a limited series of experiments conducted over a three-year period. Unfortunately the same could not be said of a lighthouse.

There is a sense of relief as Alan moves from the inscrutable effects of waves crashing randomly over rocks to something more tangible. He next considers whether the stability of an exposed lighthouse tower should depend primarily on its strength or its weight. Typically he turns to the natural world, noting that:

> the more close the analogy between nature and our works, the less difficulty we shall experience in passing from nature to art, and the more directly will our observations on natural phenomena bear upon the artificial project.

By observing the masses and shapes of rocks that have resisted destructive forces over centuries due entirely to their weight, it should be possible to make sensible design choices for man-made towers. But if it is decided to go for lightness and strength, creating a structure that has no counterpart in nature, 'it will readily be perceived that we are in a very different and less advantageous position'. Another crucial reason for favouring mass (and weight) over strength is that:

> the effect of mere inertia is constant and unchangeable in its nature; while the strength which results, even from the most judiciously disposed and well executed fixtures of a comparatively light fabric, is constantly subject to be impaired by the loosening of such fixtures, occasioned by the almost incessant tremor to which structures of this kind must be subject, from the beating of the waves. Mass, therefore, seems to be a source of stability … more in harmony with the conservative principles of nature, and unquestionably less likely to be deteriorated than the strength, which depends on the careful proportion and adjustment of parts.

Having settled for mass over strength, how much material is needed, and how should it be arranged in a tower to optimise stability? The first of these questions:

> is one of the utmost difficulty, and can be solved by experience alone, directed by that natural sagacity which Smeaton, in his account of his own thoughts on the subject, with much naivete, terms 'feelings', in contradistinction to that more accurate process of deduction which he calls 'calculation'. It is very difficult, for example, to conceive that the waves could displace a cylindric block of granite, 25 feet in diameter and 10 feet high, which would contain about 380 tons, and we almost feel that they could not do so.

Alan now attempts to bolster feelings with a calculation and, somewhat to our surprise, returns to Tom's dynamometer and the wave pressure of 4,335 pounds per square foot measured on Skerryvore in 1843–4. If this pressure were applied to the 'largest vertical section' (250 square feet) of the above block of granite, it would exert a force of 250 × 4335 = 1,083,750 pounds, or 484 tons. But of course the wave would not strike a flat vertical wall; it would meet the curved surface of a cylinder. To allow for this Alan halves the force to 242 tons, which he takes as 'the greatest force of the waves tending to displace the cylinder'.

And finally, assuming that the cylinder is completely immersed, and invoking Archimedes' Principle, its effective weight is reduced by the 140 tons of sea water it displaces, giving a value of 380 – 140 = 240 tons. Neatly (and Alan may have chosen the cylinder's dimensions to achieve this) the wave force acting sideways on the granite block is almost exactly the same as its weight, simplifying the discussion that follows.

We should pause for breath at this point because Alan's calculation relies on some crucial, and necessarily speculative, assumptions about the action of waves on a cylindrical block, and on the accuracy of Tom's dynamometer measurement which he has already questioned. It seems uncharitable, in this computer age of ours supported by more than two centuries of professional engineering, to criticise an original attempt by a great Victorian engineer to put numbers to feelings. Yet it seems that what follows may already be compromised by what has gone before.

Nevertheless Alan now asks whether the sideways force of 242 tons could overturn a granite cylinder with an effective (submersed) weight of 240 tons, or alternatively cause it to slide across the rock on which it rests.

The risk of overturning depends on the ratio between the cylinder's height and its diameter. There is a safety factor of 2.5 in the case of a cylinder 25 feet in diameter and 10 feet high, so he concludes that the wave force would need to be 600 tons, rather than 242 tons, to overturn it.

The risk of sliding depends on the amount of friction between the cylinder and its base, and Alan recalls some experiments by a Monsieur Redelet suggesting that the frictional resistance of a block of stone sliding on a chiselled floor of rock is typically 70 per cent of the block's effective weight – in this case 0.7 × 240 = 168 tons. Unfortunately this is 74 tons less than the wave force of 242 tons, but in the case of a lighthouse the excess wave force:

> would easily be neutralised by the adhesion of the mortar and by abutment of the block against the sides of the foundation pit in to which Lighthouse Towers in such exposed places are generally sunk in the solid rock.

So Alan concludes that wave forces cannot overturn or slide the block of stone, and the margins of safety are considerable. It is his most detailed attempt so far to give numerical support to lighthouse tower design. Yet the numbers rely on Tom's dynamometer experiments which 'have not been continued long enough as yet to render them available for the Engineer', and specifically on the famous reading of 4,335 pounds per square foot obtained on Skerryvore in 1843–4. In case this appears a little unsatisfactory, he has a final weapon in his armoury, the pioneering effort of John Smeaton:

> When, in addition to these considerations, we learn that the solid frustum, or lower part of the Eddystone tower, which has weathered so many storms for the last ninety years, does not greatly exceed in mass the imaginary cylindric block which I have spoken of, our confidence in the stability of the cylinder is greatly increased.

Alan has finally brought a triple brew of feelings, calculations, and professional experience to bear on the problem and it is hard to resist the thought that the last ingredient must have given him the greatest comfort. We now realise that he chose an 'imaginary cylindric block' very similar in diameter and height to the solid stub of Smeaton's Tower, which still sat resolutely on its wave-assaulted plinth, 13 miles out to sea from Plymouth Hoe.

The next major issue concerns the centre of gravity of a lighthouse tower which, for stability, should be kept as low as possible. This suggests the form of a cone with a broad base tapering to a narrow summit. But since the tower will rise well above the shock of heavy waves, the tapering effect can lessen as the height increases to make much better use of materials and provide a pleasing contour, considerations which led Smeaton to:

> the beautiful form which his genius invented for the Lighthouse Tower of the Eddystone, and which subsequent Engineers have contented themselves to copy.

However Alan's reverence for his hero is by no means unconditional, especially when it comes to Smeaton's famous analogy with an oak tree:

> Now, there is no analogy between the case of the tree and that of the Lighthouse, the tree being assaulted at the top, and the Lighthouse at the base; and although Smeaton goes on … to suppose the branches to be cut off, and water to wash round the base of the oak, it is to be feared the analogy is not thereby strengthened; as the materials composing the tree and the tower are so different, that it is impossible to imagine that the same opposing forces can be resisted by similar properties in both … and it is very singular, that, throughout his reasonings on this subject, he does not appear to have regarded those properties of the tree which he has most fitly characterized as 'its elasticity' and the 'coherence of its parts'. One is tempted to conclude that Smeaton had, in the first place, reasoned quite soundly, and arrived by a perfectly legitimate process at his true conclusion; and that it was only in the vain attempt to justify these conclusions to others, and convey to them conceptions which a large class of minds can never receive, that he has misrepresented his own mode of reasoning.

Alan's knockout blow, uncharacteristically brutal, demolishes Smeaton's argument. The timber trunk of an oak tree, bending in the wind due to assault at the top, is emphatically not the same as a stone tower, resisting movement with its mass and assaulted by waves at the bottom. And although Smeaton's tower is undoubtedly beautiful, he has done himself no favours by appealing to popular imagination with a superficially attractive but false analogy. Fortunately Alan's demolition job is slightly softened by his final comment that 'with the single exception of the allusion to the oak, he (Smeaton) has discussed the question throughout in a masterly style'.

To summarise, the general form of a lighthouse, typified by the Eddystone, should be generated by the revolution of a suitable curve about the tower's axis, gradually approaching the vertical towards the top. But the precise form of the curve cannot be established until other dimensions of the tower – its height, base diameter, and top diameter – have been fixed.

Height is extremely important because providing the light is strong enough it is the height that limits a lighthouse's range. Alan does not explain this point in his *Account*, merely noting that the height can be determined 'by means of the known relations which subsist between the spheroidicity of the earth, the effects of atmospheric refraction, and the height required for an object which is to be seen from a given distance'. We can, in fact, explain it quite easily using the famous theorem of Pythagoras that in a right-angle triangle 'the square on the hypotenuse equals the sum of the squares on the other two sides'. This is illustrated in the accompanying sketch.

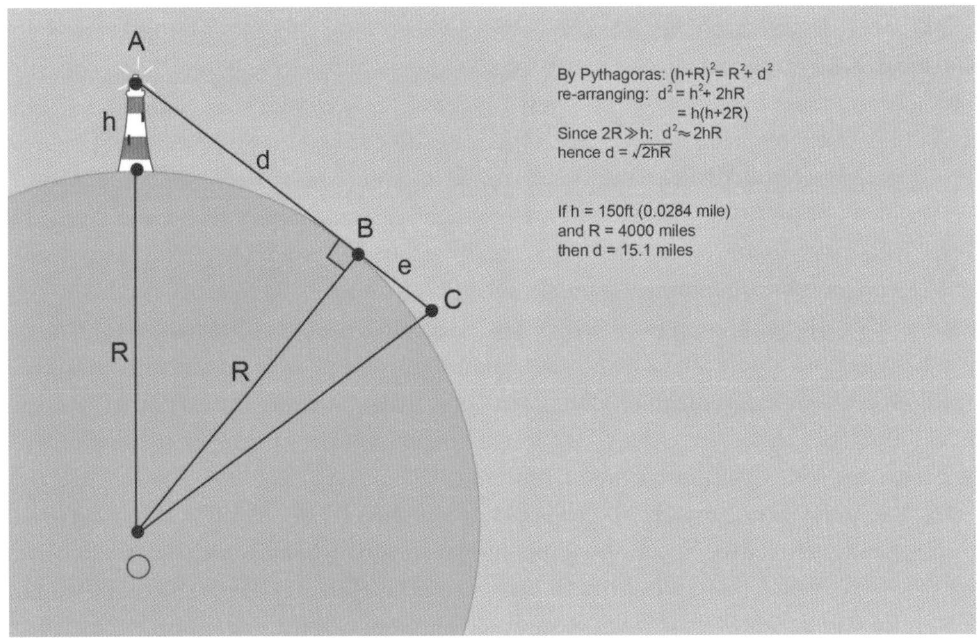

The relationship between the height of a lighthouse and its range.

The lighthouse A of height *h* on a spherical earth of radius *R* sends its light towards the horizon at B. The maximum range at which it may be seen by a mariner on the sea surface is *d*. Since the line of sight is a tangent to the earth, we have a right-angle triangle AOB, and it is simple to calculate the range using Pythagoras' theorem. For example if *h* is 150 feet (0.0284 mile) and we take the radius of the earth as 4,000 miles, the range is 15.1 miles.

Of course, a mariner observing the light is not actually on the sea surface – unless the ship has gone down and he or she is in a lifejacket. Their eye is far more likely to be at some point, C, above it. This gives us another right-angle triangle, BOC, and the range is extended by the amount *e*, which may be calculated in exactly the same way. For example, if the mariner's eye is 25 feet above the sea, the range increases by 6.2 miles to 21.3 miles.

There is one more complicating factor mentioned by Alan – the 'effects of atmospheric refraction'. Basically, the density and refractive index of the earth's atmosphere vary with height due to a temperature gradient. The index is normally greatest near the surface, causing a light beam to be refracted slightly downward and increasing the distance to the horizon. In standard atmospheric conditions the difference is about 8 per cent, but over the sea it is often considerably greater, perhaps as much as 20 per cent, which would increase the above range of 15.1 miles to 18.1 miles.

The range of a lighthouse may also be limited by the strength of the light. There is no point building a high tower if the light is too weak to be visible, even in clear weather. The Eddystone was initially lit by 'common candles' and the Bell Rock by reflector lamps, but Alan is aiming at a more sophisticated lens system. His lighthouse must protect ships not only from Skerryvore itself, but from other rocks spread across the extensive reef including the dreaded Boinshly and Bo-rhua, and some outliers 3 miles to seaward. Taking all this into account he opts for a masonry tower 138.5 feet high, elevating the light to 150 feet above high water at spring tides and giving a range 'of at least 18 miles'. Equivalent figures for the Eddystone and Bell Rock are about 12 and 15 miles respectively.

The decision has major consequences. His father trumped Smeaton's 68-foot Eddystone tower with a 100-footer on the Bell Rock, needing twice as much stone; but Alan wants 138.5 feet of masonry, doubling the weight yet again to about 4,300 tons. He needs 100 per cent more granite, almost all to be shipped to Tiree and Skerryvore from a distant quarry and fashioned with enormous skill to demanding tolerances. Although his lighthouse is nearly 40 per cent higher than his father's, thanks to Pythagoras and the earth's curvature it gives only 20 per cent more range. It is a tough call, but if it goes ahead the lighthouse will be majestic in scale as well as elegant in outline.

At this point it is tempting to reopen Alan's discussion about the tension between 'feelings' and 'calculations'. The increase in range given by a higher tower is easy to calculate, but the human benefits are subjective. How many mariners' lives will be saved over the next century or two by a few more miles of range? How often will the extra range be nullified by poor visibility? How many workmen will be lost or severely injured as they struggle skyward to lay the extra stone courses? There are no clear answers to such questions and it finally comes down to feelings and engineering instinct. Alan is no doubt aware of the imponderables and his decision to recommend a 138.5-foot tower to the Commissioners seems all the braver because of them.

There are more decisions to be made before the tower's shape can be finalised, often interacting and sometimes pulling in opposite directions – a classic design engineer's challenge. Alan decides that his isolated lighthouse needs about 13,000 cubic feet of 'void space' to accommodate the keepers in reasonable comfort and provide stowage for provisions and stores. Given the limited area of Skerryvore free of the dangerous gullies and fissures that run 'much farther into the Rock than might at first be imagined', he settles on a base diameter of 42 feet, identical to the Bell Rock. Of the tower's 138.5 feet, 18 are allocated to the decorative capital supporting the light room and 26 to the solid base, or frustum. He decides on a 12-foot diameter void in the remaining 94.5 feet, sufficient to house men and stores and to receive the lantern and apparatus. And finally, he considers that the walls just below the capital:

> should not be less than 2 feet thick, as necessary to give due solidity and strength to the walls, and prevent, by the breadth of the joints, the percolation through the walls of the water which might be furiously dashed against them in storms … .

He can now define the curve line connecting the top and bottom radii of the tower's shaft which, allowing for the 18 feet allocated to the capital, are separated by a vertical distance of 120 feet. Here comes the decision that will largely determine the tower's reputation for elegance as well as technical design, and Alan gives it a great deal of thought. He is not prepared simply to scale up the beautiful but intuitive curve of Smeaton, but favours something to reflect the purity of mathematics:

> For this purpose I tried four different curves, the Parabola, Logarithmic, Hyperbola, and Conchoid, figures of which, upon the same scale, will be found in Plate No IV, with the position of the centre of gravity, which was carefully calculated, marked on each. The logarithmic curve I at once rejected, from its too near approach to a conic frustum, and the excessive thickness of the walls which such a figure would produce … The parabolic form displeased my eye by the too rapid change of its slope near the base … and of the last two curves I preferred the former, which gives the most advantageous arrangement of materials, in regard to stability, of all the four forms.

It is another design decision based on the tension between 'feelings' and 'calculation'. He rejects the logarithmic form because of its excessive thickness of walls, the parabolic because it displeases his eye, and the conchoid because of its slightly less advantageous arrangement of materials. The hyperbola takes the prize.

The 'slight superiority of the hyperbolic over any of the other forms' is quantified in a table giving the volume M of each tower in cubic feet, and the distance G of its centre of gravity above the base, for a tower of height H. Alan calculates two parameters. The first, equal to H/G, is used as a measure of the tower's stability – for a tower of given height, the lower the centre of gravity the better. The second, a rather curious quantity he calls the 'economic advantage', is inversely proportional to the product GM – for a tower of given height, the lower the centre of gravity and the smaller the volume the better (a stable tower can be achieved with a relatively small amount of stone). Unsurprisingly Alan's favourite, the hyperbola, comes top of the class in both categories, albeit by extremely small margins. It is hard to fault his judgement even though its scientific and economic justification seems

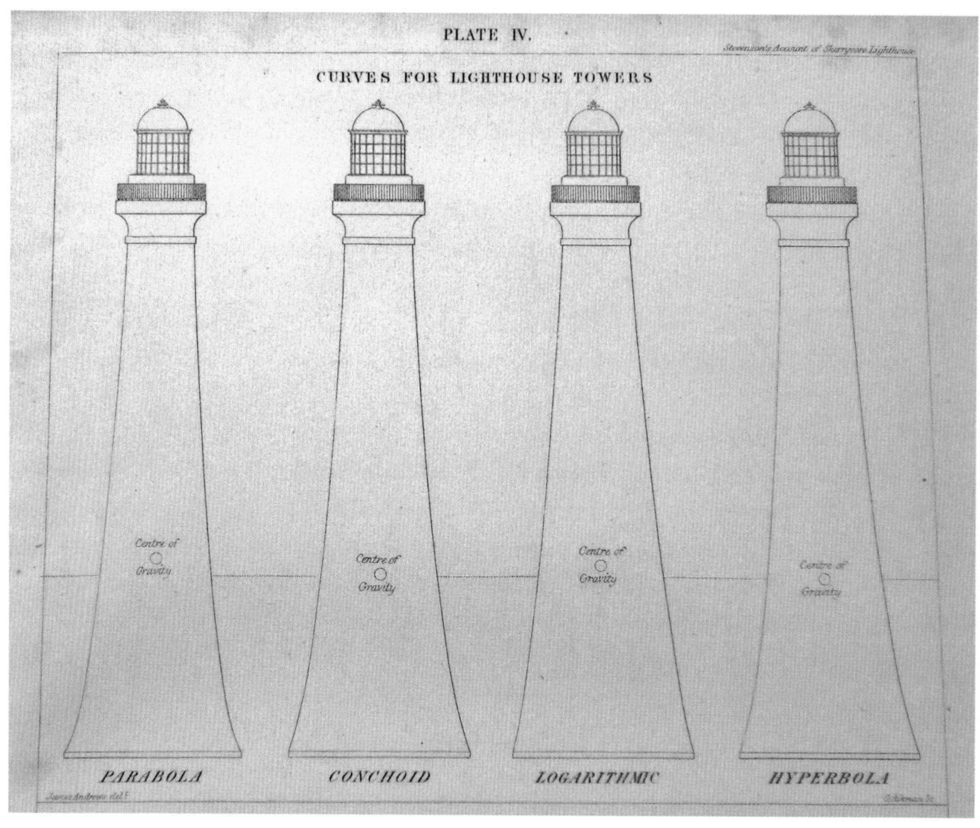

Four candidate curves for a lighthouse shaft with a height of 120 feet, base diameter of 42 feet, and top diameter of 16 feet. Alan's favourite, the hyperbola, is on the right.

slender. It is surely as much to do with beauty as utility. And his tower has been praised ever since for its graceful outline.

Aware of the pivotal decisions he has made, Alan now gives a technical summary:

> The shaft of the Skerryvore Pillar, accordingly, is a solid, generated by the revolution of a rectangular hyperbola about its asymptote as a vertical axis. Its exact height is 120.25 feet, and its diameter at the base 42 feet, and at the top 16 feet. The ordinates of the curve, at every foot of the height of the column, were carefully determined in feet to three places of decimals; and the Appendix contains a tabular view of the co-ordinates from which the working drawings were made at full size. The first 26 feet of height is a solid frustum, containing about 27,110 cubic feet, and weighing about 1990 tons. Immediately above this level the walls are 9.58 feet thick, whence they gradually decrease throughout the whole height of the shaft, until at the belt they are reduced to 2 feet in thickness.

Using the cool beauty of mathematics Alan has calculated the radius of the shaft foot by foot above the base, to within one hundredth of an inch. It is hard to imagine 120.5 feet of drawings 'at full size', but there it is: his masons and artificers have a superb guide, and very little excuse for error as they fashion over 4,000 tons of granite blocks and assemble them like a gigantic three-dimensional jigsaw puzzle.

We can almost hear a sigh of relief as the engineer celebrates the grandeur and uniqueness of his design with a set of woodcut illustrations, inserted in the main text of the *Account* rather than relegated to the end with the other plates:

> It may, perhaps, be not uninteresting to the reader to examine the woodcuts which shew, on one scale, the elevations of the Lighthouses of the Eddystone, the Bell Rock, and the Skerryvore, and exhibit the level of their foundations in relation to high water. They will also serve to give some idea of the proportionate masses of the three buildings. The position of the centre of gravity, as calculated from measurements of the solids, is also marked by a round black dot on each tower; and in the table following, I have given the cubic contents of each of these towers, the height of the centre of gravity above the base and the ratio of that quantity to the height of the tower.

The comparisons with Eddystone and Bell Rock speak volumes about his ambition not only to light Skerryvore, but to do so with a tower of stunning proportions built with 58,580 cubic feet of granite. About the only advantage, from the point of view of the engineer who must erect it, is that the base remains above high water mark at spring tides, unlike the Eddystone and especially the Bell Rock. During the early stages of construction the workforce can avoid fleeing the foundation pit and lower courses on every rising tide.

Alan brings the chapter to a close with a description of tower features that are, to put it mildly, controversial:

> I come now to notice the few subordinate points in which the design of the Skerryvore Tower may be regarded as differing from those of the Eddystone and the Bell Rock. In glancing at the contrasted figures of the three buildings, it will be at once observed that the outline of the Skerryvore approaches more nearly to that of a conic frustum than the other two. To the adoption of this form, various considerations induced me; and these I shall very briefly detail.

First, he repeats his intention to increase the wall thickness at the top, improving safety and reducing tremor in violent weather. This gives a curve with slightly 'less concavity' than the Bell Rock and reduces the excess of the bottom radius over the top radius from 13.5 to 13 feet. He next divides the shaft's height of 120.5 feet by 13 feet, giving a quotient of 9.27 to represent the average slope of the walls away from the perpendicular. And now for a key point:

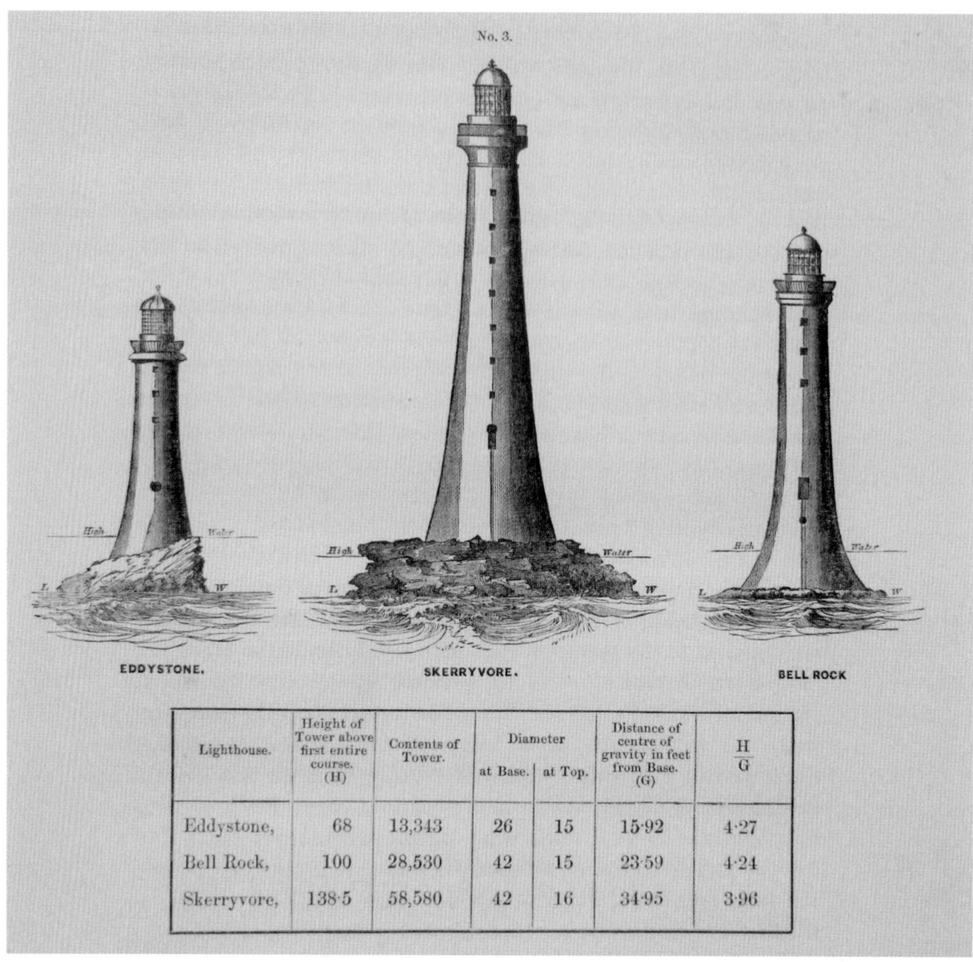

Lighthouse.	Height of Tower above first entire course. (H)	Contents of Tower.	Diameter		Distance of centre of gravity in feet from Base. (G)	$\frac{H}{G}$
			at Base.	at Top.		
Eddystone,	68	13,343	26	15	15·92	4·27
Bell Rock,	100	28,530	42	15	23·59	4·24
Skerryvore,	138·5	58,580	42	16	34·95	3·96

Eddystone, Bell Rock, and Skerryvore compared.

> There can be little doubt that the more nearly we approach to the perpendicular, the more fully do the stones at the base receive the effect of the pressure of the superincumbent mass as a means of retaining them in their places, and the more perfectly does this pressure act as a bond of union among the parts of the tower.

The 'bond of union' is enhanced when the weight of a high tower is concentrated onto a relatively small base and distributed evenly between the inner and outer stones. He notes that the quotient of 9.27 contrasts with values of 7.53 for the Eddystone and 6.59 for the Bell Rock which, 'without, however, venturing to enunciate any general law', he feels must indicate the relative strengths of the bond. And now for the author's equivalent of a naval broadside:

> This consideration seems too important to be entirely overlooked; and I conceive that, by following out this view, I have been enabled

> to depart with perfect safety from the intricate and elaborate work required for the connection of the materials by means of dovetailing and joggling, which the adoption of a more concave outline (in which the vertical pressure could not have been so advantageously transmitted to the outer stones of the base), would perhaps have rendered advisable.

With a single sentence Alan challenges the extensive dovetailing and joggling pioneered by Smeaton and copied by Rennie and his father – design features widely credited with keeping the Eddystone firmly anchored in the English Channel for nearly a century and the Bell Rock in the North Sea since Alan was a boy. Arguing his case carefully, but unsupported by any definitive theory, the designer of Skerryvore pushes his boat out into the North Atlantic 'with perfect safety'. He is relying on weight, bolstered by the thought that his 4,300-ton tower is twice as heavy as his father's but rests on a base of identical diameter. Furthermore, it is less flared out at the base, giving a more equable weight distribution between the inner and outer stones. But his courageous decision is as much about feelings as calculation and, some would say, risks disaster.

Alan is fully aware of the implied criticism and in fairness admits that the Bell Rock tower:

> is covered to the depth of 15 feet at spring tides, and that this principle of vertical pressure could not have been safely appealed to during the whole time which intervened between the commencement of the building and the attainment of a height sufficient to render it available, which, in a Tower having so great a part submerged, was of necessity much prolonged. The stones were thus exposed to the full effect of heavy seas, at all levels, during two entire winters, and could not therefore have been safely left, without being kept together by numerous ties and dovetails. It also seemed important, in designing that Tower, with reference to the rise of tide, to give its lower part a sloping form, as the least likely to obstruct the free passage of the waves. The outer stones of the lower courses were also selected of unusual length inwards, so as to bring them more under the influence of the vertical pressure of the upper wall.

Alan's design did not avoid dovetails and stone joggles altogether. Although he dispensed with them *between* courses in the lower part of the tower, individual courses were stabilised during the early stages of the work by simple 'diamond' joggles inserted in the outer ring of stones, and temporarily united to those above and below by wooden trenails as in the Eddystone and Bell Rock. He also used some more sophisticated connections towards the top of the tower where the walls were thinnest, and tied the walls to the various floors of the building with dovetails. But the great majority of Skerryvore's 97 stone courses were essentially kept in place by the weight bearing down from above.

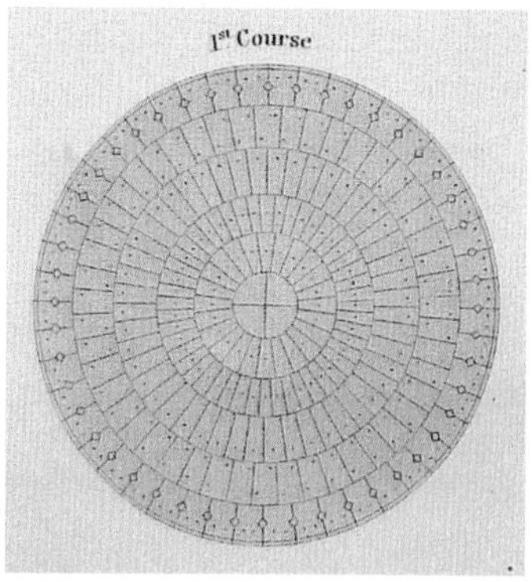

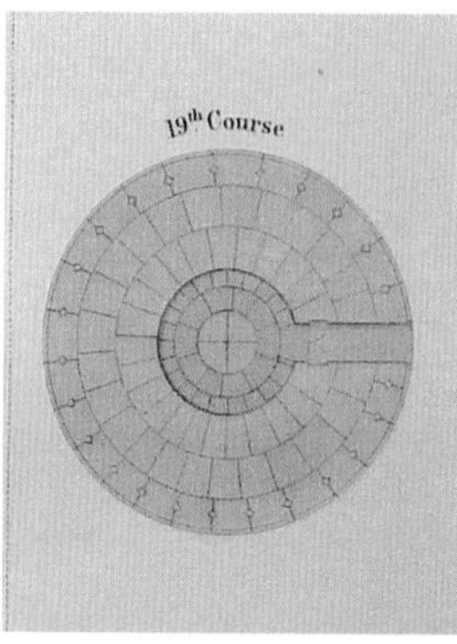

The first course of Alan's design, showing the diamond joggles and trenails; and the 19th course, the lowest in 'the void', showing the diamond joggles. There are no dovetails.

By contrast, his father used no less than 338 joggles, 4,065 trenails, 6,329 pairs of wooden wedges, and a huge number of dovetails in the first 38 courses of the Bell Rock.

Alan's connection scheme was undoubtedly simpler than those of the Bell Rock and Eddystone. It saved his masons much 'intricate and elaborate work' and the Commissioners a lot of money. As a statement of his design philosophy on the construction of lighthouse towers it could hardly be bolder, but we may imagine the ageing Robert Stevenson casting a sceptical eye over his son's plans and, just possibly, John Rennie stirring in his grave.

1838: A FALTERING START

> The hazardous nature of the anchorage, and the consequent difficulty of mooring a vessel in the neighbourhood of the Skerryvore Rock, induced me, from the first, to consider it as a matter of great importance, even at a large expenditure of time and money, to erect some temporary dwelling on the Rock for the accommodation of the people engaged in the work, with the view of rendering the operations less dependent on the state of the sea, which varied with every wind. So important, indeed, did this object appear to me, that I was at times apt to look upon it as an indispensable step towards ultimate success.

As it turned out, Alan's determination to erect a 'temporary dwelling' on Skerryvore dominated the construction work of 1838 and, following a disaster towards the end of the year, a considerable part of 1839.

The temporary barrack erected on the Bell Rock by his father 30 years before had proved of enormous benefit. Rather than being forced off the rock by rising tides and crammed into sailing boats that pitched sickeningly in North Sea swells, Robert Stevenson and his men could shelter and sleep in relative safety and comfort above the rock itself. The barrack had survived the storms of five winters and Alan was happy to adopt the same design, which consisted of:

> an open framework of six logs, about 47 feet long and 13 inches square, assembled in such a manner as to form by their union a hexagonal pyramid, on the top of which rested a wooden turret; the whole erection rising to the height of about 60 feet above the rock. This pyramidal framework was strongly trussed and tied; and, being open at the lower part, offered little resistance to the waves … The small space which the turret afforded was, with the utmost economy of room, divided into three storeys, of which the lower was entirely taken up by the kitchen and the bread-store … The next storey was subdivided into two chambers, of which one was appropriated to the foreman of the works and the landing master, while the other was set apart for myself; and the top storey, which was surmounted by a small lantern and ventilator, formed a barrack room, capable of containing thirty people.

Valuable though such a barrack might be, it is not clear how 30 people could be 'contained' in a room 7 feet high and 13 feet across with a floor area of 4 square feet per person, especially during a week-long Hebridean storm; nor is there any mention of washing or toilet facilities.

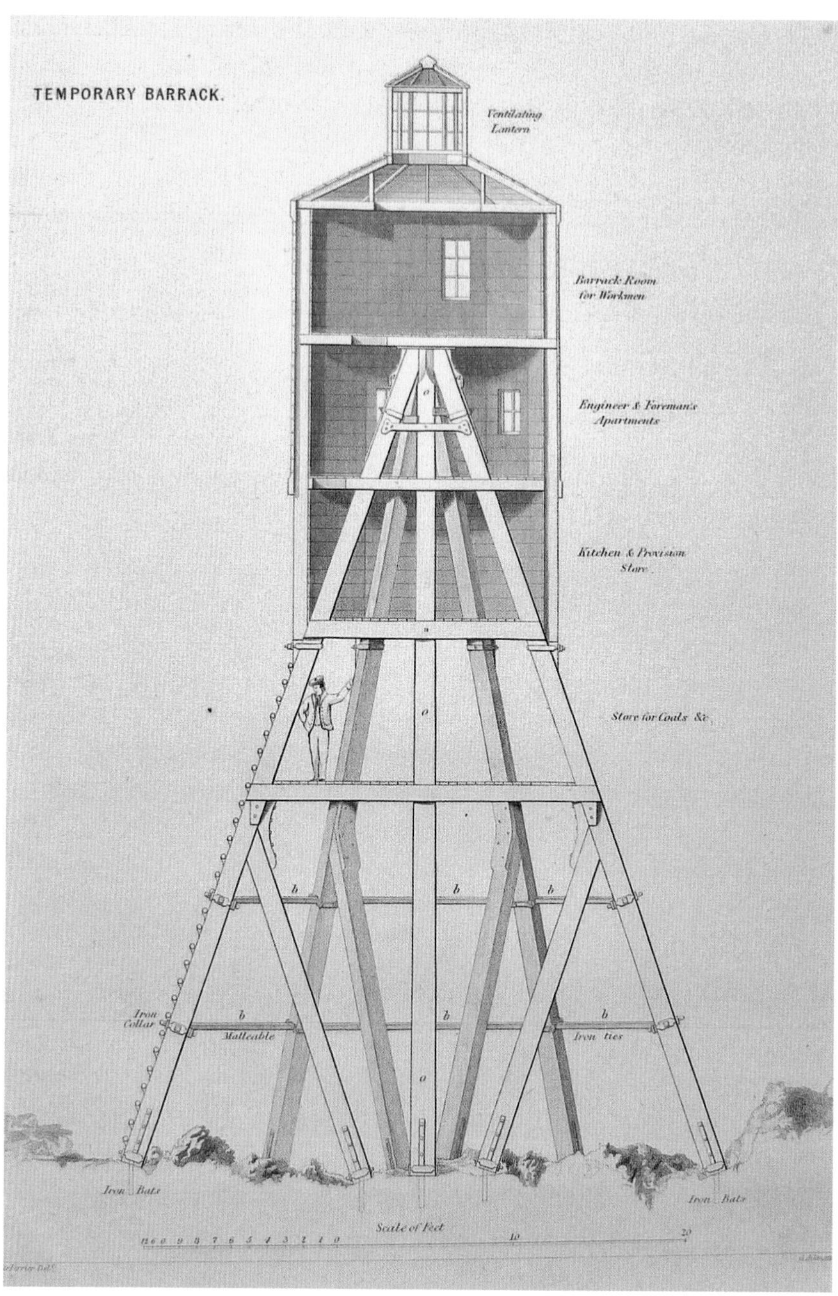

Design for the temporary barrack.

At first sight Alan's sanguine acceptance of a building that depended for its survival on strength rather than weight seems surprising. Having extolled the virtues of weight for his lighthouse tower, he now settles for a wooden building on stilts. However the barrack was always intended as a temporary structure, so perhaps there was nothing to worry about. He obtained estimates from several carpenters in Greenock on the Clyde and chose a Mr

John Fleming to make the new 'log-house' and 'fit it up' in his workyard before shipment to Tiree. He also ordered a 'large assortment of quarriers' and masons' tools of every kind; and many cranes, crabs, anchors, mooring buoys and other implements, according to detailed specifications and drawings'.

Transport was the next major concern. Steam navigation had become common in the 30 years since Robert had transferred men and materials to the Bell Rock in sailing boats, and Alan wanted to make best use of the short working season with a vessel that could overcome the vagaries of wind and tide. He considered no less than 24 used steamers offered for sale from various ports around the British Isles, but decided that the Skerryvore service needed a robust new vessel. A shipbuilder in Leith, Edinburgh, was contracted to build a 150-tonner powered by two 30 hp steam engines. Unfortunately it could not be delivered before 1839 and in the meantime he had to struggle with the elderly *Pharos,* the 36-ton sailing vessel used by his father for the Bell Rock service, notorious for inducing sea sickness and a long way past its prime. It was:

> the only regular shipping attendance we possessed during this first season; and the inconvenience arising from her heavy pitching, was, to landsmen, by no means the least evil to be endured. But the frequent loss of opportunities, of which we might easily have availed ourselves, if we had possessed the command of steam-power, and the danger and difficulty of managing a sailing vessel in the foul ground near the Rock, and between it and Tyree, were, perhaps, even more felt by the seamen than by the landsmen … .

Another pressing issue was the hire of skilled workers to start cutting and dressing stones for the tower and, even more urgently, to excavate holes in the Skerryvore rock to secure the barrack's wooden legs. Tiree could not oblige so Alan's foreman, Mr James Scott, was sent to Aberdeen to select 30 masons and stone-cutters, 12 quarriers, and several smiths for two years of 'certain employment'. They were promised good wages, a provision store at Hynish, and barrack accommodation or lodgings ashore with free cooking facilities.

By early May 1838 Alan was able to inspect the completed barrack in Greenock and check the other equipment and supplies, including two portable forges for use on the rock. A vessel was chartered to carry coal to fuel-starved Tiree. He was becoming involved in a multitude of arrangements far removed from lighthouse construction and was feeling the pressure:

> In providing the means of efficiently carrying on so many complicated operations in a situation so difficult and remote, it is impossible, even with the greatest foresight, to avoid omissions; while delay of a most injurious kind may result from very trivial wants. Even the omission of a handful of sand, or a piece of clay, might effectually stop for a season the progress of plans, in the maturing of which hundreds of

pounds had been expended. Accordingly, although I had bestowed all the forethought which I could give to the various details of the preparation for the season (of which I found it absolutely indispensable to be personally aware, even to the extent of the cooking dishes), new wants were continually springing up … .

It was undoubtedly tough. Tiree could supply very little apart from sand and the engineer had better things to do than order cooking dishes. He was entering six years of extraordinary attention to detail involving thousands of decisions which he might have hoped to delegate. But, as so often, the leader found that if the organisation was to be done reliably he had better do it himself. It must have been a great relief to board the *Pharos* at Tobermory in Mull on 23 June, loaded with supplies for the season's operations, and head out towards Tiree.

One of the most striking aspects of Alan's *Account* is the extent to which his plans and schedules were disturbed by contrary winds and heavy seas, especially before the new steamer was delivered. Time after time the *Pharos* was unable to leave port, or was driven in the wrong direction. A huge amount of time was spent waiting for winds to change or waves to subside, and every opportunity offered by favourable weather had to be grasped – even if it meant arriving or departing at an ungodly hour.

On this first occasion, the *Pharos* arrived off Hynish Point at three in the morning on Sunday 24 June and disembarked the passengers in rough seas. Nothing moved on Tiree on Sundays so they delayed sailing for Skerryvore until six the following morning, but a 'foul wind' followed by calm meant that it took 21 hours to cover the 12 miles to the rock. They managed to lay a mooring a quarter of a mile off, in 13 fathoms of water, but Alan's attempt to land on the rock from a rowing boat was thwarted by heavy surf. After 'hanging on their oars' for nearly an hour, hoping for a lull, the crew gave up and with the wind still freshening the master of the *Pharos* decided to reef his mainsail, set the first jib, and run downwind for Mull. The next day was spent sheltering and it was 28 June before Alan was able to set foot on Skerryvore – four frustrating days after leaving Tobermory. It was by no means the worst example of weather-induced disruption and delay.

Whenever Alan was forced to mark time he made every effort to use it. While stranded at Hynish on 24 June he and his foreman inspected the new range of buildings being erected, including stores and barracks for up to 100 men, a magazine for gunpowder, and 100 feet of landing pier. On 27 June, with *Pharos* sheltering in Mull's North Bay, he managed to inspect the 'almost inexhaustible supply of flesh-coloured granite' on the Ross of Mull, opposite the island of Iona, and decided that this was where most of the stone for his lighthouse would come from. He made notes on suitable sites for a landing place, new quarries, and barracks for a Mull workforce.

Alan's first landing on Skerryvore was terrifying:

a landing could not be attempted till mid-day, when I went with Mr Macurich in the boat, and with some difficulty contrived to spring on the Rock, after which the boat returned to the vessel for the rest

> of the party. While left alone on this sea-beaten Rock, on which I had landed with so much difficulty, and as I watched the waves, of which every succeeding one seemed to rise higher than the last, the idea was for a few minutes forcibly impressed on my mind, that it might, probably, be found impracticable to remove me from the Rock, and I could not avoid indulging in those unaccountable fancies which lead men to speculate with something like pleasure upon the horrors of their seemingly impending fate. These reflections were rendered more impressive by the thought that many human beings must have perished amongst those rocks.

By the time the rowing boat returned from the *Pharos* with the rest of the party Alan had regained his composure and was planning the erection of the temporary barrack. It was a day of 'great bustle and interest', spent chalking out and marking with paint locations for the barrack, tower, and heavy equipment. They spent four hours on the rock, much to the annoyance of seals and innumerable sea-fowl driven from their favourite haunts; but the birds got their own back when the sun, rising high in the midsummer sky, magnified the stench of droppings and cast-off feathers. After a heavy day the party sailed for Tiree and reached Hynish the next morning.

A day later, 30 June, the *Pharos* set sail for Greenock with Alan on board. It took nearly a month to assemble and check all the machinery and materials ordered in Greenock and Glasgow and load them on two hired vessels, the *New Leven* and the *Mary Clark*. The *Pharos*, also laden, was too low in the water for the risky passage round the Mull of Kintyre so had to return to Tiree via the Crinan Canal, reaching Hynish on 4 August. Three days later Alan and his foreman Mr George Middlemiss, accompanied by 4 carpenters, 16 masons and quarriers, and a smith, caught a favourable northerly wind to Skerryvore and put in 'what may be called our first good day's work on the Rock'. It must have been a huge relief after all the preparations, setbacks, and delays; however:

> Our work on the rock was by no means easy, as we had to erect shear-poles and fix crabs for landing the materials, and to lash every article that was landed, with great care down to ring-bolts on the Rock … as yet we had no tramways on which wheeled trucks could be moved, and the transport by hand of heavy materials over so irregular and slippery a surface was attended with considerable danger … but in spite of all the fatigue and privation attending a day's work on this unsheltered Rock, the landsmen were for the most part sorry to exchange it for the ship, which rolled so heavily as to leave few free from sea-sickness, and to deprive most of the workmen of sleep at night, even after their unusually great exertions during the day.

Alan and his men made a trial boring of a hole for one of the iron stanchions, or 'bats', by which the timber legs of the barrack were to be secured in the rock. It was their first skirmish with the gneiss of Skerryvore, a stone 'considerably more difficult to bore than even the granite of Aberdeenshire', and it took an hour to fashion a hole 3.5 inches in diameter and just 3 inches deep. The day's work done, the men had the greatest difficulty getting back to the *Pharos* in two small boats which were tossed around so violently that they risked being thrown right up onto the tender's deck.

The next day, 8 August, the chartered vessel *New Leven* arrived with heavy materials and made fast to the *Pharos*. The day was largely spent landing the principal timbers and iron fixtures for the barrack, and lashing them safely to the rock.

As night approached a 'great and sudden stillness of the air', coupled with a moon which rose 'red and fiery', alerted the sailors among them to an impending gale; and sure enough, by midnight a stiff south-easterly forced the master of the *New Leven* to cut loose and run. Unfortunately the seaman on watch on the *Pharos* was asleep and it was two in the morning before Mr Macurich roused Alan. They were moored just a few hundred yards to windward of the rock in a rising gale, the tide was strong, and the sea increasingly violent:

> We soon set sail, but in vain tried to weather the sunk rock Bo-Rhua, whose large black mass (after having imagined ourselves past it) we discovered encircled by a wreath of white foam within less than a cable's length of us ... A more anxious night I have never spent; there being upwards of thirty people on board, with the prospect, during several hours, of the vessel striking every minute. And here I must award due praise to Mr Macurich for the coolness and intrepidity which he on this occasion displayed, and the calmness with which he gave his orders to the crew ... We had no sooner cleared the sunk rocks already alluded to, than we were in fear of the great reef of Boinshly, and the heavy seas which were breaking over the foul ground all round it. In this way we spent a night of almost uninterrupted anxiety ....

They had survived the triple terrors of Skerryvore, Bo-Rhua, and Boinshly, and after a terrifying run back to Hynish:

> were not sorry to set foot even on the wild shores of Tyree; and I trust there were none who did not gratefully acknowledge the protecting care of Almighty God, in preserving us through such peril.

They were then blessed with six days of fine weather.

Alan's *Account* now focuses on the major task of the season – the erection of the hexagonal pyramid of timbers to support the turret of the barrack. The engineer is again in his element, intrigued by the technicalities of securing a wooden structure to the uneven surface of a wave-assaulted rock. The work proved nerve-racking and dangerous but by 11 September, a mere five weeks after the first 'good day's work', the timbers were in place.

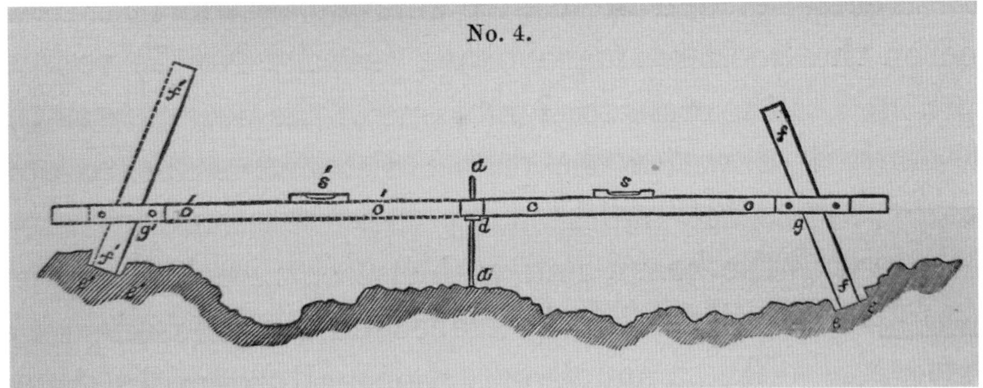

Alan's method for precise location of the barrack's timber legs.

His description of the pyramid is far more detailed than his father's for the similar structure on the Bell Rock. The six wooden legs had to be precisely located, allowing for the ups and downs of the rocky surface and meeting precisely at the top. Needless to say, Alan had come well prepared with ideas and materials. After some initial preparation of the rock by blasting with small shots, an iron rod (*aa*) was driven vertically into the rock at the desired centre of the building, as shown in the accompanying drawing. The rod served as a pivot for a wooden frame (*gg*) some 40 feet long which was kept horizontal with the aid of two spirit levels (*ss*). Two short timber frames (*ff*), with the same 13 × 13 inch cross-section as the legs, were inserted into grooves at the ends of the main frame at precisely the required angle, and could slide up and down to make contact with the rock. This showed the masons exactly where, and at what angle, the rock should be dressed to provide accurate footings; and by measuring the amount of the short frames (*ff*) extending below the main frame (*gg*), it also indicated the precise length to which each leg should be trimmed so as to meet its neighbours accurately at the top. Finally, holes 3.5 inches in diameter by no less than 24 inches deep were hand-bored into the rock to receive the iron stanchions, or 'bats', which secured the legs to the rock. The process was repeated for the other legs by rotating the main frame horizontally through 60 degrees, and similar footings were prepared for six inner braces. The work kept 12 men busy for four days.

With the structure assembled, a plumb line was suspended from the top and, to Alan's huge satisfaction, settled within half an inch of the central rod (*aa*) more than 40 feet below. This he attributed to the 'intelligence and zeal' of Mr George Middlemiss, his foreman carpenter, who had overseen the construction and trial assembly of the pyramid in Greenock. In spite of great difficulties raising and manoeuvring legs that weighed over half a ton each, the structure had taken just six days to re-erect on Skerryvore.

They had enjoyed a week of fine weather and had nothing to complain about apart from long hours of toil on the rock and sea sickness aboard the *Pharos*. However Alan admits that the men's lifestyle and food was 'somewhat singular'. Generally they landed on the rock at four in the morning and breakfasted at eight:

> at which time the boat arrived with large pitchers of tea, bags of biscuit, and canteens of beef … work again resumed, till about two o'clock, which hour brought the dinner, differing in its materials from breakfast only in the addition of a thick pottage of vegetables, and the substitution of beer for tea. Dinner occupied no longer time than breakfast, and like it, was succeeded by another season of toil, which lasted until eight and sometimes till nine o'clock, when it was so dark that we could scarcely scramble to the boats … Once on the deck of the little tender and the boats hoisted in, the materials of breakfast were again produced under the name of supper; but the heaving of the vessel damped the animation which attended the meals on the Rock, and destroyed the appetite of the men, who, with few exceptions, were so little seaworthy as to prefer messing on the Rock even during rain, to facing the closeness of the forecastle.

The season's work on Skerryvore was largely completed by 18 August, and Alan needed only a few more days to brace the pyramid with horizontal struts and tie equipment down against winter storms. But it was not to be. The North Atlantic, which had been leaving them in peace for up to 16 working hours a day, decided to reassert its authority and play havoc with the schedule. They were forced back to Hynish, and then Tobermory, by a gale; thence back to Hynish again, where they had just about enough time to load some materials and a highly unappetising 'salted sheep' before running for shelter off the west coast of Coll. They returned to Skerryvore on the evening of 30 August, but the following day was made miserable by a thick drizzle, ugly waves that drove them off the rock for several hours, and a 'most distressing' sea that forced them back to Hynish yet again. And so it continued for another 12 days, as though Skerryvore was determined to show Alan and his men, before they returned to mainland comforts, that it would never yield without a fight. At one point a 'marrot' (probably a guillemot) perched on the vessel's side, 'much fatigued and evidently desirous to get on board; but the sailors, from some superstitious dread, would not admit the poor bird'.

Fortunately Alan had already planned to delay mounting the accommodation turret on top of the wooden pyramid until the following year. So during brief respites from the awful weather the men managed to complete the season's work and were ready to leave on 11 September. Alan, in a gesture of solidarity, had a chest containing biscuits and fresh water lashed to the top of the pyramid in case any shipwrecked mariners found themselves on Skerryvore against their will. He had one final ambition:

> Before leaving the rock, I climbed to the top of the pyramid from which for the first time I got a birds-eye view of the various shoals which the stormy state of the sea so well disclosed; and my elevation above the Rock itself decreased the apparent elevation of the rugged ledge so much, that it seemed to me as if each successive wave must sweep right over its surface, and carry us all before it into the wide

> Atlantic. So loud was the roaring of the wind among the timbers of the barrack, and so hoarse the clamour of the waves, that I could not hear the voices of the men below; and I, with difficulty, occasionally caught the sharp tinkle of the hammers on the Rock. When I looked back upon the works of the season, upon our difficulties, and, I must add, dangers, and the small result of our exertions – for we had only been 165 hours at work on the Rock between the 7th of August and the 11th September – I could see that, in good truth, there were many difficulties before us; but there was also much cause for thankfulness, in the many escapes we had made.

A rousing three cheers rang out as they left Skerryvore. Alan spent the next few days on Tiree making arrangements for the 1839 season before returning to the mainland, with absolutely no premonition of disaster.

The bombshell arrived on 12 November in the form of a letter from Mr William Hogben, the clerk and storekeeper at Hynish, saying that the timber pyramid had been totally destroyed during a south-westerly gale on the night of 3 November. According to the locals, the Atlantic swell had been the worst for 16 years. Two days later another letter brought news that a party led by foreman Mr Charles Barclay had managed to land on the rock, where they found a desolate scene: most of the timbers carried away; iron stanchions broken and partly drawn from their holes in the rock; an anvil thrown 8 yards; cranes and the smith's forge nowhere to be seen, and other items damaged or missing.

On the day the second letter arrived Alan received instructions from the Northern Lighthouse Board to hire a steamer at Glasgow and proceed immediately to the rock. He left Greenock at midnight on 14 November in the steamer *Tobermory* accompanied by Mr Macurich, trusted master of the *Pharos*, and reached Hynish 36 hours later. They picked up Mr Charles Barclay and sailed for Skerryvore in heavy seas. Although there was not the slightest chance of landing, they managed to get near enough to inspect 'the melancholy remains of our labours' and found conditions very much as Mr Hogben had described. A miserable return to Hynish was accompanied by 'all the gloom of a stormy night … depressed by mingled disappointment and sad forebodings'. A strong gale prevented them from landing until seven the next morning on an island blanketed in snow, and when they sailed again the deck was covered by sea-sick and utterly dispirited passengers. Hearts sank as the vessel scraped a rock off Hynish point, but it struggled to Oban without further incident.

It is hard to imagine Alan's state of mind at this point. He had faced enormous physical danger getting to know Skerryvore in all its moods and had erected the timber pyramid against huge odds. There must have been moments in the following days when he was tempted to throw the towel far into the North Atlantic and admit that Skerryvore's sceptics had been right all along – it was an impossible project. Perhaps the Northern Lighthouse Board would be wise to withdraw before wasting more of its precious funds. Yet there were overriding personal considerations. He had been dreaming of Skerryvore ever since agreeing to join his father in the lighthouse business – and, after all, it was Robert's design for the pyramid that had failed, not his own.

In any case his *Account* gives no hint of doubt. Putting his engineer's hat firmly back on, Alan first considers, then rejects, the suggestion that the pyramid was destroyed by shipwreck flotsam that came ashore on Tiree two days after the disaster. He settles instead for an engineering failure of the horizontal braces, which allowed the main timbers 'more liberty for play and tremor' and gradually dislodged the fixtures at the apex:

> The moment this dismemberment occurred, the beams would be free to work their own destruction; and the enormous leverage which they exerted, when dashed to and fro by the breakers, would soon snap the iron stancheons [*sic*] at the base, and throw all loose to the waves.

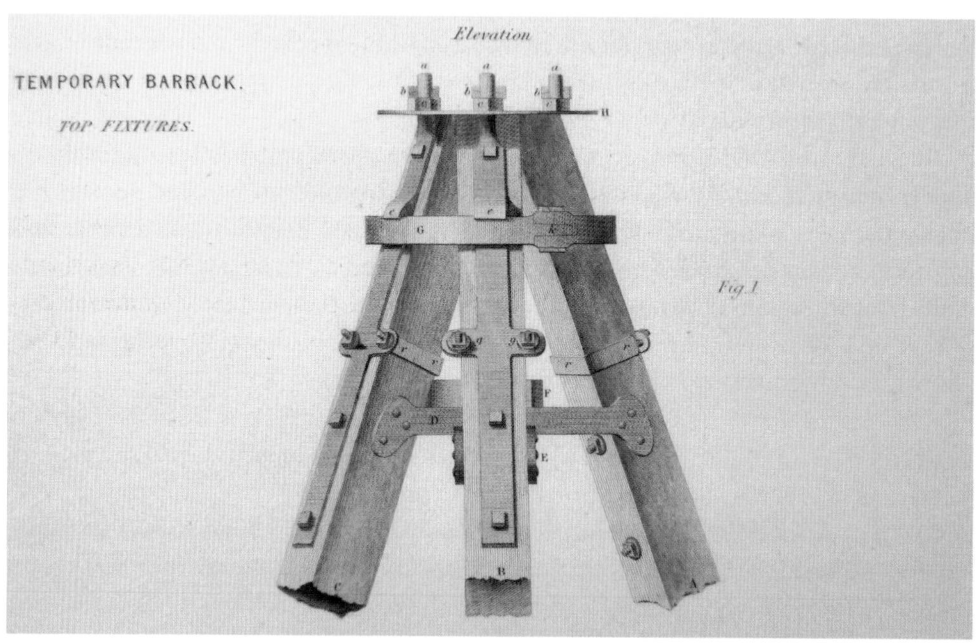

The apex of the redesigned pyramid with its elaborate iron fixtures.

Heartened by the Commissioners' assurance of continued support, Alan set about re-designing the pyramid. He strengthened the apex with iron stanchions (*aaa*), bolted to the main legs and united at the top by an iron centreplate (H). Further fixtures bound the six legs together into a compact unit to resist successive wave impacts. Finally, he braced the lower levels of the structure with horizontal struts of malleable iron in place of timber – and presumably crossed his fingers.

In spite of all the troubles on Skerryvore, the works at Hynish continued apace. About 30 men were employed building the onshore accommodation and workshops. A large room was fitted with a stone floor for laying out full-scale drawings of the tower's courses and making timber moulds for dressing the stones. Outside, a masonry platform was constructed for assembling and checking each completed course, the pier was extended, and stores were built for tools, materials, and coal for the desperately needed steamer. A large food store was established on the second floor of one of the buildings:

> to preserve the provisions from injury by damp and to secure them from the inroads of the needy Celts and from innumerable rats which overrun that part of Tyree … The chief articles served out in the store were meal, molasses, sugar, coffee, tea, tobacco and butter … the inhospitable nature of the country, and the remoteness of Tyree from the ordinary steam-boat traffic, made the adoption of some such plan unavoidable.

In spite of many requests, local labourers were not allowed to purchase items from the food store, 'except in cases where dearths (which are of frequent occurrence in the island) rendered the call irresistible'. By the late 1830s Tiree was becoming desperately overpopulated and underfed.

So what had actually been achieved in the first brief season on Skerryvore? Sadly, the answer must be rather little. Alan first reached the rock on 28 June but spent most of July in Greenock and Glasgow. Weather disruptions meant that the working party put in only 165 hours on the rock between early August and mid-September. And finally, the pyramid of the temporary barrack had disappeared. About the best that can be said for the summer of 1838 is that it gave them plenty of insight into what they would have to tackle in 1839.

1839: Solid progress

A striking difference between Alan's account of Skerryvore and his father's earlier one of the Bell Rock is the amount of space devoted to engineering principles. Robert Stevenson, man of action, tends to gloss over theory and focus on day-to-day progress; the highly educated and contemplative Alan prefers to weigh ideas and report his findings carefully. As we have seen, the third chapter of his *Account* is a mini-treatise on the construction of lighthouse towers. Much of Chapter V: 'Operations of 1839' is devoted to a detailed discussion of the building material that dominated his professional life: Scottish granite.

Trial quarrying at Hynish in 1837 had yielded so little stone of decent size and quality that Alan took every opportunity to visit other sites. He inspected, and rejected as unsuitable, old red sandstone from quarries near Oban and porphyry from Inveraray before settling on pink granite from the Ross of Mull. After a winter on the mainland he sailed from Greenock to Mull on 19 April 1839, accompanied by the components of a new temporary barrack, and reached North Bay after a 'tedious passage of 6 days'. The next day dawned far more cheerfully:

> we had the satisfaction of seeing the steamer, the *Skerryvore* (by which name she was specially set apart for the service of the works), arrive in the bay with a party of masons and quarriers, who had been appointed to meet us in order to begin the work.

It is surprising he makes so little of the new arrival. The elderly *Pharos* had tried his patience and the stomachs of his workmen terribly in 1838 and he must have been delighted to swap sail for steam. The 150-ton *Skerryvore,* with her gleaming twin 30 hp engines, would transform the business of transporting men and materials between Mull, Hynish, and the rock.

Work on Mull started impressively. It took Mr Charles Stewart, the foreman builder, and his team just a week to erect the barrack's turret as temporary accommodation, and three months to complete various storehouses, permanent dwellings for up to 40 men, and a 40-foot landing wharf. A quarry was opened in a steep hillside and linked to the wharf below by wheeled trucks running on an inclined plane. Alan was highly enthusiastic about the quarry and its yield:

> a large mass of most beautiful granite, whose reddish colour is said to have given the name of Ross to that part of Mull ... I have never seen a granite quarry of equally great resources, as regards both the quantity and the quality of the material produced. The rock in general yielded easily before well-directed shots, and was separated into large masses, ca-

> pable of being advantageously cut, with little loss of material, in to shapely blocks, by means of wedges, which work remarkably well in that rock.

Unlike Tiree gneiss, a quarrier's nightmare, the pink granite of Mull was a dream. Huge rocks could be dislodged with a single gunpowder charge, and in just over a year 26 quarriers, 3 labourers and 2 smiths fashioned them into 4,300 rough blocks weighing between ¾ and 2½ tons.

Alan now interrupts his chronological account of 1839 with a mini-treatise on the mining and dressing of granite, and for a very good reason:

> As I am not aware that any professional work contains a detailed description of the quarrying of granite, some observations on that subject may not be unacceptable in this place; and I therefore propose, at the risk of appearing somewhat prolix, to give a pretty minute account of the mode of opening and working a granite quarry, more especially as practised by us at North Bay.

He explains the full range of quarrying tools and skills, including boring holes for gunpowder charges, best practice to safeguard the lives of fireman and quarriers, techniques for dislodging large masses of rock from a steep cliff, and the use of iron wedges to split them into smaller blocks. The fine dressing of stone was equally important:

> and as no writer with whom I am acquainted has given any account of the mode now practised of dressing granite, I hope I shall be excused for attempting, in this place, to give some idea of the method employed by the masons of Aberdeenshire, whose skill in that department of workmanship is well known both in our own and in other countries.

He follows with a highly expert account of dressing granite to the unusual patterns and fine tolerances required by his lighthouse tower.

Up to 70 men were employed in the dressing shed at Hynish, fashioning every stone to mesh perfectly with its neighbours before shipment to Skerryvore. Starting with hammers weighing up to 40 pounds and finishing with a wide variety of lighter tools, the masons used timber moulds to achieve a remarkable degree of accuracy. The time spent on a stone depended greatly on its complexity and size, averaging about 80 man-hours; but a record 320 man-hours was needed for a large, dovetailed, centre stone for one of the tower's upper floors.

After ten days on Mull Alan sailed for Skerryvore, determined to make the most of a new summer season on the rock. He was about to swap cooperative pink granite for stubborn black gneiss, and was painfully aware of what awaited him:

> The excavation of the foundation of the Lighthouse Tower was the first operation which engaged my attention at Skerryvore Rock, at the

> beginning of the season of 1839. It was commenced on the 6th of May, and was continued up to the last hour of our remaining on the Rock, on the 3rd of September. A more unpromising prospect of success in any work than that which presented itself at the commencement of our labours, I can scarcely conceive. The great irregularity of the surface, and the extraordinary hardness and unworkable nature of the material, together with the want of room on the Rock, greatly added to the other difficulties and delays, which could not fail, even under the most favourable circumstances, to attend the excavation of a foundation-pit on a rock at the distance of 12 miles from the land. The Rock, as already noticed, is a hard and tough gneiss, and required about four times as much labour and steel for boring as are generally consumed in boring the Aberdeenshire granite.

The foundation pit was to be the same size as his father's on the Bell Rock. Fortunately it would not flood on every rising tide and need pumping out, but the gneiss of Skerryvore was vastly more difficult to work than Bell Rock sandstone. Traversed by 'numerous veins and bands inclined at various angles', it would threaten the stability of any building placed upon it. Alan decided to start with a rough levelling of the topsy-turvy surface to reveal any serious imperfections and, together with his foreman Mr Charles Barclay, took personal responsibility for charging bore holes with gunpowder and firing the shots. The lethal menace of flying splinters was reduced by covering the holes with nets and mats, and the debris had to be dumped in deep water around the rock to prevent it being thrown about by storm waves and causing damage. The work occupied 30 men for 102 days; 246 gunpowder shots were fired, and 2,000 tons of unwanted gneiss removed.

Alan now had a clear view of the surface's imperfections and, after 'repeated examinations of all the veins and fissures', marked out a circle 42 feet in diameter. He wanted to avoid laying partial stone courses as Smeaton had done on the Eddystone, preferring to start with a perfectly level pit. Using great caution and minimal gunpowder, the gneiss was excavated to a uniform depth of 15 inches and finished with a clean vertical edge around the 132-foot circumference. The painstaking work kept 20 masons busy for 168 days in 1839, and would require a further 49 days in 1840.

Alan's *Account* now moves on to the thorny question of the temporary barrack which had given him so much anguish at the end of 1838. It was as vital as ever to the success of the project, but he is clearly out of love with it and affords it little space:

> The minute details given in my account of the destruction of the first Barrack, have entirely superseded the need for any particular description of the fitting up of the second Barrack on the Rock; and I shall therefore confine myself to a brief notice of the work in the chronological order in which it occurred.

The working party started erecting the barrack on 1 July when the final shot for the foundation pit had been fired, and a fortnight later the redesigned and strengthened timber pyramid stood proud – and hopefully secure – above the rock of Skerryvore. The turret house, which had been shipped from Mull to Hynish, was transported piecemeal to the rock as required and mounted on the pyramid to complete a work 'of great difficulty and anxiety'.

There was also sorrow. Mr George Middlemiss, Alan's trusted lieutenant and foreman carpenter who had overseen assembly of the barrack at Greenock, died suddenly of a 'paralysis of the heart' two weeks after completing his labours on Skerryvore. There seems little doubt that, in his case at least, the temporary barrack had the last word.

They were not quite finished with gunpowder on the rock. The natural creek used by the boats to disembark men and materials was jagged and dangerous and needed blasting out to make a decent landing place and wharf. Alan pressed technology ancient and modern into service:

> After two entire days spent in cutting with a sickle, mounted on a long pole, the thick cover of gigantic sea-weed, which hid the true form of the rock from view, we were able to mark out the line of the wharf; and after all the mines were bored and charged and the tide had risen, and every one had retired from the spot, the whole were fired at the same instant, by means of the galvanic battery, to the great amazement and even terror of some of the native boatmen, who were obviously much puzzled to trace the mysterious links which connected the drawing of a string at the distance of about a hundred yards, with a low murmur, like distant thunder, and a sudden commotion of the waters in the landing place, which boiled up, and then belched forth a dense cloud of smoke; nor was their surprise lessened, when they saw that it had been followed by a large rent in the rock; for so effectually had the simultaneous firing of the mines done its work, that a flat face for a quay had been cleared in a moment, and little remained to be done, to give the appearance of a regular wharf and to fit it for the approach of a stone lighter, except attaching wooden fenders and a trap ladder.

Other important tasks included clearing a level line for a 50-yard permanent iron railway to convey stones and other materials from the landing place to the tower, and installation of a large cast-iron tank for fresh water. A severe August gale intervened and gave everyone a lot of worry, but the barrack held firm. So did Alan and Mr Heddle, master of the new steamer *Skerryvore,* when threatened by a few disgruntled seamen who demanded higher wages, spread fear among the landlubbers, and spoke of mutiny. Alan recalled his father's similar troubles on the Bell Rock in 1810 and decided on a similar course of action: the troublemakers were deposited unceremoniously back at Hynish and replaced by 'native boatmen'. The episode was not to be repeated, although there were rumblings from time to time over wages and working conditions, plus the occasional booze-fuelled threat of violence.

1839 certainly produced its fair share of challenges. Loss of life or limb still threatened whenever men attempted to land or leave in rough weather, but unlike 1838, which had ended in disaster, the second season on Skerryvore was marked by solid progress. The many vessels that passed within a few miles of the rock left Alan more than ever convinced that his lighthouse was sorely needed:

> It often happened that for several days successively, not fewer than five or six vessels of large size, both outward and homeward bound, were visible at distances varying from 3 to 6 miles from the Rock; and much anxiety was often felt by us for the safety of those vessels, several of which approached so near the outlying rocks as to keep us for some time in the most painful suspense. On two occasions, more especially, I was about to direct the steam to be raised, in order that the Skerryvore tender might be sent to warn the masters of vessels of their danger, or if too late for that, to afford the assistance in case of accident.

Alan left Skerryvore's temporary barrack and foundation pit for the winter comforts of Edinburgh on 3 September, quietly confident of eventual success. As the working party sailed away to Hynish:

> every heart was full of rejoicing, and many cordial expressions of gratitude to our Almighty Protector were uttered in deep whispers by the more seriously disposed men, whose number bore a goodly proportion to our whole band.

1840: Stones to Skerryvore

> The building of the Tower was commenced on the 4th July; but it was not till the 7th that the ceremony of laying the foundation-stone was performed by His Grace the Duke of Argyll, who, as Proprietor of the adjacent Island of Tyree, took a great interest in the success of the works, and on that day visited the Skerryvore with the Duchess of Argyll, the Marquis of Lorne, lady Emma Campbell, and a party of friends, in the *Toward Castle* steamer. On that occasion His Grace expressed himself much pleased with the works and kindly left with me a donation of L.10 for the workmen.

Summer on Tiree is by no means free of occasional storms and heavy Atlantic swells. Although the day chosen for the visit was reasonably calm the duchess panicked a few miles out of Hynish and the party had to return. The disappointed duke and his factor, accompanied by Alan, set out once more for the rock and the foundation stone was laid with less ceremony than planned, fortified by a modest 'three cheers and a glass'. Alan was far too tactful to record the embarrassing reality of the occasion – or his own irritation at the loss of valuable time.

The three main centres of activity – Hynish, Ross of Mull, and Skerryvore – worked at full stretch in 1840. Alan's attention and concern switched constantly between them, sorting out the multitude of problems that inevitably arose in a project of such complexity. In his *Account*, rather than arrange Chapter V1, 'Operations of 1840', in chronological order, he avoids to-and-fro confusion by describing 'the various departments separately, beginning with the workyard at Hynish'.

The Hynish workforce had been reduced to about 50 over the preceding winter; 27 masons dressed granite blocks for the lighthouse; 9 carpenters were fully occupied making timber moulds, oak trenails, and handles for the masons' and quarriers' tools; and 18 quarriers and labourers were given the unenviable task through the sombre days of a Hebridean winter of claiming gneiss from the Tiree quarries for use in the new pier at Hynish.

In April a reinforcement of 37 highly skilled masons arrived from Aberdeen and the clatter of steel on stone in the dressing shed became deafening. As spring turned to summer some of the masons were deployed to Skerryvore to complete the foundation pit. The skilled workers remaining at Hynish helped land large quantities of granite arriving at the pier from Mull, using cranes and heavy lifting tackle that were beyond the competence of local labourers. By the time Alan left Tiree for Edinburgh at the end of October, no less than 20,000 cubic feet of dressed stone had been checked and assembled into courses. Roughly half had been shipped to Skerryvore and laid to form the first six courses of his tower.

Hynish pier was in constant demand in 1840 and could barely take the strain:

> During the whole of the summer, the traffic at the pier at Hynish was so great in landing materials from the Mull quarries, and in

> shipping stones for the Rock, that much inconvenience was felt from want of room. Nearly 4000 tons were shipped and discharged at the quay, independently altogether of the ballasting of each vessel which discharged at the pier, and the receiving, storing, and finally supplying coals to the steamer, which formed no inconsiderable item of the labour. Every exertion was made to extend the pier, so soon as the works at the Rock were closed for the season and the stone trade with Mull had ceased; and by great perseverance on the part of Mr James Scott, the foreman of the workyard, whom I always found ready, night and day, to second and even to anticipate my wishes in regard to the progress of the works, an additional length of 36 feet was added to the berthage of the quay before the winter had set in.

On 27 March a party sailed from Hynish to Skerryvore after some heavy gales to inspect the state of the barrack. There were no obvious problems with the building itself, but a 5-ton block of stone, loosened by blasting operations the previous year, had been carried right across the foundation pit in a storm and dumped against one of the main beams of the timber pyramid. Clearly, it had to go – and go fast. The men started breaking it up for dispersal and, 'allured by the smoothness of the sea', decided to leave their boat in the creek, stay overnight in the barrack, and finish the job the next day. But the wind got up. They had no fire, light, or bedding, and almost no food; and the cold dark night proved so full of anxiety and discomfort that several of them 'did not afterwards much affect the Rock as a residence even in summer'. Nature had issued a stern reprimand.

Alan first visited the rock on 30 April, and was hugely relieved to find the barrack in good order after its long winter vigil. Seabirds had covered it with malodorous messages, protective paintwork had bleached, seaweed had settled and ironwork had coated with rust; but his redesigned timber pyramid was intact. When they forced open the door of the turret and unbarred the windows:

> the interior was in some places so dry, that the greater part of the biscuits which we had left the year before, as a store for shipwrecked seamen who might find their way to the Rock, although some were wet and pulpy on the side nearest the outer walls, admitted of being dried, and when a little toasted at the fire, were palatable enough to hungry men, so that, in fact, we consumed the greater part of that stock before we entered on our new supply.

Accommodation in the turret was certainly preferable to being tossed around all night on a boisterous sea, but the first few months proved extremely uncomfortable. They had almost no furniture or facilities, heavy spray and rain forced their way through cracks in the walls, and it was sometimes extremely cold. On one occasion Alan and his men:

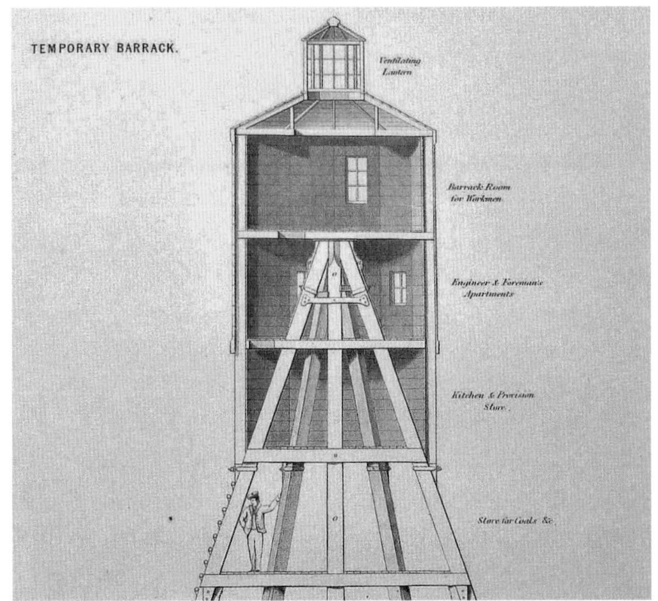

Turret accommodation (from bottom to top): platform for storage of coals etc; kitchen and provision store; engineer's and foreman's apartments; barrack room for workmen; ventilating lantern.

> were fourteen days without communication with the shore or the steamer; and during the greater part of that time we saw nothing but white fields of foam as far as the eye could reach, and heard nothing but the whistling of the wind and the thunder of the waves, which were at times so loud as to make it almost impossible to hear any one speak.

They spent most of the fortnight in their beds, unable to descend to the rock, listening to the howling winds and beating waves which 'occasionally made the house tremble in a startling manner'. One particularly violent wave produced cries of terror from the men in the top apartment, convinced that they had all been washed into the sea. When the storm finally subsided they signalled to Hynish that a landing was possible, and greeted the arrival of *Skerryvore* 'with the greatest joy'. They had just one day's provisions left.

Immediately below the wooden turret an open platform or gallery was provided for storing coal, tool chests, beef and beer casks, and other items that could not be safely left on the rock itself. The floor timbers had to be removed at the end of the season to allow free passage of waves during winter storms. Next came the kitchen and provision store, six-sided and 12 feet in diameter, lined with tin and 'sheet-lead' [*sic*] for fire protection, and furnished with food cupboards, lockers, and cooking facilities for up to 40 men.

The next storey was divided into two apartments. One was shared by Mr Macurich, former master of the *Pharos* and newly appointed as landing superintendent, and Mr Charles Stewart. The other was allocated to Alan. Together the rooms occupied a narrow 12-sided space twisted crazily around the central pyramid, with sloping walls on the inside. Although Alan appreciated his own semi-private apartment he had few illusions about its generosity:

> The half of that space constituted my apartment, which, I think, would be generally pronounced not over commodious; and when it is added that it contained my bed, desk, chair and table, and a stock of groceries, it will readily be imagined I had little room to spare for myself. So much attention was paid to economy of space, that the recesses of the pyramid formed by the meeting of the beams were boarded over and made into cupboards; while my cot, or framed hammock (which, during the night, rested upon brackets which could be folded close to the wall when not required), was, during the day, hoisted by pullies [*sic*] to the roof of the apartment, so as to leave me as much space to move about in as a prisoner could expect.

'Rabbit hutch' might have been a more honest description of the apartment that was to form the engineer's bedroom, study, and home from home over the next three summers.

The top floor of the turret was allocated to the workmen – rarely less than 30 in number during the 1840 season. How they fitted into a space 12 feet in diameter and 7 feet high, especially during a prolonged storm, defies the imagination; at full capacity the volume per man was less than a tenth of that provided by a modern Scottish prison cell. Presumably they had no option but to while away the time in their bunk beds, stacked four high around the perimeter. And how about washing, shaving, and toilet facilities? Fortunately the turret was topped with a generous 'ventilating lantern', but even so there must have been many occasions when the atmosphere, emotional as well as physical, became well-nigh intolerable. Alan admits that in fine weather many of the men preferred 'taking a nap on the Rock, with the clear blue sky for a canopy'.

The working day started with a call at half-past three in the morning, and in good weather might continue until nine in the evening, with half-hour breaks for breakfast, dinner, and supper. Exhaustion was common:

> Such protracted exertion produced a continual drowsiness, and almost every one who sat down fell fast asleep. I have myself repeatedly fallen asleep in the middle of breakfast or dinner, and have not infrequently awakened, pen in hand, with a half-written word on the paper!

Yet Alan never complains, least of all about himself. Even in such daunting conditions he finds time to revel in the grandeur of nature all around him, and reflect upon his place within it.

The main task during the first half of the season was to finish the foundation pit for the tower, and it kept 30 men busy for up to 14 hours a day. In many places the gneiss was so hard that steel picks made almost no impression. Three strokes of a tool were enough to blunt it and the smiths were kept extremely busy at their forges. Alan was so concerned by small splinters of rock flying off the tools that he sent urgently to Glasgow for fencing masks to protect the men's faces.

As the work proceeded nothing was more begrudged than the occasional loss of half a day baling out the pit after it had been filled by a heavy sea. But skill and perseverance were finally rewarded by a masonry masterpiece:

> a foundation pit so level and so fairly wrought throughout the whole area of a circle 42 feet in diameter, as to present to the view the appearance of a gigantic basin of variegated marble; and so much pleased were the workmen themselves with the result of their protracted toil, that many of them expressed serious regret that the foundation must soon be covered up so, as (we trusted), never to be seen again.

Another of Alan's major concerns in 1840 was the safe delivery of dressed stones from Hynish. It was extremely risky to unload stones weighing a ton or more from a conventional vessel heaving with every wave, and in bad weather the crane might even be toppled before a stone could be cleared from the hold. His father had faced the same problem on the Bell Rock and solved it with a special design of dumb barge, or lighter, decked all over to carry up to 20 tons of stone, with nothing below except empty casks for emergency buoyancy. Alan had ordered four similar vessels from Edinburgh and Dumbarton, and used the steamer to tow them back and forth between Hynish and the rock, where they were moored in the creek for unloading.

A trial landing was made on 20 June. The steamer and lighters were gaudily decorated with flags, and successful unloading of the first stone was greeted with a salvo shot and a wee dram. It was, however, just the start of a long and often bitter struggle:

> The landing service throughout the whole progress of the works was one of much difficulty and anxiety and many narrow escapes were made; but it was managed with great prudence, and at the same time with unremitting energy, by Messrs Macurich and Heddle, in their several departments, both ashore and afloat. On many occasions the men who steered the lighters ran great risks; and it was often found necessary to lash them to the rails, to prevent their being thrown overboard by the sudden bounds of the vessels, or being carried away by the weight of water which swept their decks as they were towed through a heavy sea. Sometimes, also, we were forced, owing to the rush of the sea into the creek, which threatened to lift the vessels on the top of the Rock, to draw out the loaded lighters from the wharf without landing a single stone, after they had been towed through a stormy passage of 13 miles … During the whole season, however, in the course of landing 800 tons of masonry on the Rock, too often in that dangerous manner, none of the dressed stones received any great damage, nor was any other injury of importance sustained.

The building of the tower began in earnest once the foundation stone had been laid by the Duke of Argyll on 7 July. Two cranes were used, one to forward materials from the landing place, the other placed in the centre of the tower to lay the stones. Alan is full of praise for the Hynish masons, whose accuracy meant that no additional dressing was needed. A mortar mixed from equal parts of Aberdda lime and Pozzolano earth, identical to that used by Smeaton on the Eddystone, gave remarkably good adhesion and freedom from leaks.

The first course was laid using a 'wooden trainer', and the accuracy of subsequent ones was checked with 'plumb templets' (vertical templates) whose inner faces were arcs of Alan's famous hyperbolic curve. With practice, Mr Charles Stewart and his men found they could set as many as 85 blocks in a single day and by the end of the season six courses, the first three of Tiree gneiss, the rest of Mull granite, had been completed 'in a most perfect manner'. The stub now rose 8 feet 2 inches above the foundation pit and contained 10,780 cubic feet of stone, almost as much as Smeaton's complete Eddystone tower.

The unyielding gneiss of Tiree was hardly denser than the workable granite of Mull. Had the Skerryvore tower been built entirely of gneiss it would have weighed about 4,308 tons; entirely of granite, 4,252 tons. The stability of the tower, which depended on its total weight, was therefore virtually unaffected by Alan's decision to use pink granite from the fourth course upwards and as the season continued his enthusiasm for it was undiminished:

> The Hynish stone is harder, and susceptible of finer workmanship, and perhaps its most perfect blocks are more durable; but it requires much more labour in dressing than the Mull granite, which is more homogeneous in its structure and is not intersected by hard veins, like those which occur in the gneiss of Tyree. There is good reason also for concluding that the Mull stone is sufficiently durable, because it contains but a small portion of micaceous matter, and in its texture closely resembles some of the blocks of St Oran's chapel in the neighbouring Island of Iona, which have resisted the action of the weather, it is believed, for more than 600 years and still retain the marks left by the tools of the workmen.

As a final comment on 1840, Alan cannot resist comparing the excellent progress at Hynish with a depressing local scene:

> the desolation and misery of the surrounding hamlets of Tyree seemed to enhance the satisfaction of looking on our small colony, where about 150 souls were collected in a neat quadrangle of cleanly houses, conspicuous by their chimnies [*sic*] and windows amongst the hovels of the poor Hebrideans, who generally make no outlet for the smoke in their gloomy dwellings, but permit it to escape by the

St Oran's Chapel on Iona, five miles across the sea from the Ross of Mull, contains many blocks of pink granite (Wikipedia).

> doors. The regular meals and comfortable lodgings and the cleanly and energetic habits of the Lowland workmen, whose days were spent in toil and their evenings, most generally, in the sober recreations of reading and singing, formed a cheering contrast to the listless, dispirited, and squalid look of the poor Celts, who have none of the comforts of civilised life and are equally ignorant of the value of time and the pleasures of activity.

Sad, but probably true. And presumably the typical reaction of a highly educated lowland Scot, brought up in Edinburgh's thriving New Town, to the life style of semi-destitute Hebridean islanders.

1841: Onward and upward

As 1841 dawned, Alan must have wondered what additional burdens would be placed on his shoulders, and whether the tower could continue upward without major setbacks. His experience so far had been extremely mixed. The disastrous loss of the barrack's timber pyramid at the end of 1838 had reduced 1839 to little more than a catch-up year, but 1840 saw its replacement installed, the foundation pit completed, and the first few stone courses laid. Given reasonable weather, a reliable steamer, and a share of good fortune, it should be possible to maintain momentum.

More than 80 masons were employed during the year, crafting stones to an eighth-inch tolerance in the dressing shed at Hynish, loading and unloading cargoes at the wharf, and keeping an eye on local labourers who 'partly from incapacity and partly from excessive indolence could not be trusted for a moment to themselves'. The excellent Mr James Scott, workyard foreman, oversaw the dressing of 38,000 cubic feet of pink granite into myriad shapes and sizes, including 70 dovetailed floorstones, 'a work of great nicety'. The stones were numbered according to a detailed schedule, checked and then stacked in the workyard, and shipped out to the rock as required.

Rough spring weather delayed the first landing on the rock until 13 May. Alan was delighted to find the barrack undamaged and the stub of the tower undisturbed. A full workforce arrived a week later in the steamer but foul weather forced most of them straight back to Hynish, leaving Alan to risk landing with a small party of masons armed with tools to clean up the debris of winter. The first real crisis of the year was immediate, and personal:

> One of the masons took alarmingly ill soon after the steamer was too far off for a signal, and suffered so acutely during the whole night, that his piercing cries in the spasms which accompanied his disorder, combined with the howling of a strong north-wester and the incessant lash of the waves, deprived the whole party of sleep during the first night.

The steamer reappeared two days later with the main party, and carried the unfortunate man back to Hynish. Alan does not mention him again, nor the attention he received from Dr Campbell, the resident surgeon.

The first cargo of stones was landed on 25 May. The next day a 34-year-old crane, used on the Bell Rock by Alan's father, was erected on top of the stub and 'the more cheering operations of mixing the mortar and of setting stones were begun'. Progress was so impressive that they completed the solid part of the masonry, up to the 18th course, by early July – exactly a year after the Duke of Argyll had laid the foundation stone.

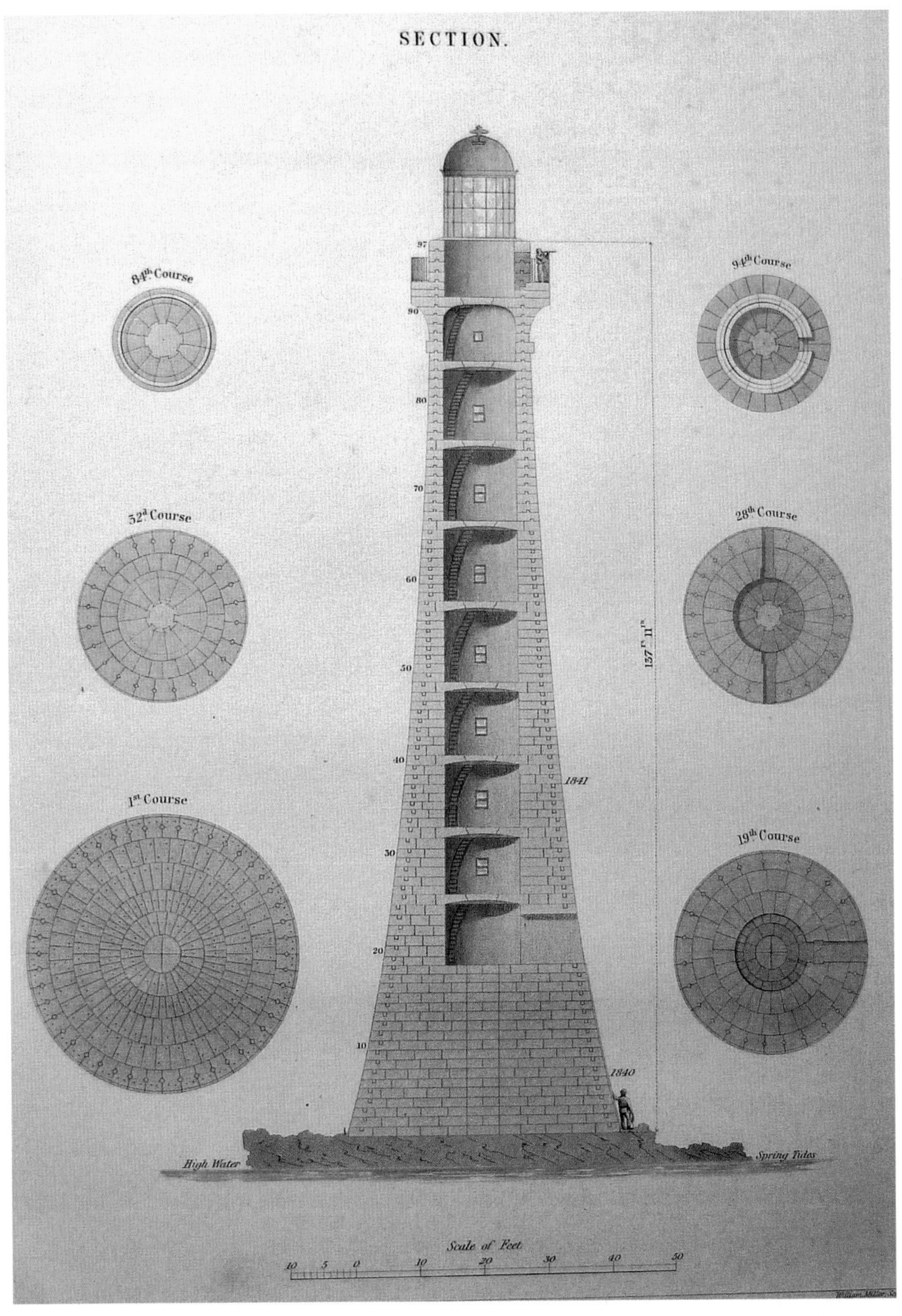

A section through the lighthouse. Course numbers and levels reached at the end of 1840 and 1841 are shown against the tower.

Initially they used one jib-crane on the rock and another on the tower to raise and position the stones. But when the masonry reached a height of 15 feet they switched to 50-foot timber 'shear-poles' erected at the side of the tower, similar to those used by Smeaton on the Eddystone. Stones could now be raised 40 feet above the rock and transferred to the tower crane. As the building grew even higher they used lifting tackle and horizontal beams or 'needles' projecting from the windows to complete the lift, as Robert had done on the Bell Rock.

Unfortunately the first part of the 1841 season continued cold and miserable, with showers of sleet and heavy spray:

> which dashed round us in all directions, to the great discomfort of the poor masons, whose apartments did not admit of a large wardrobe, while they had not the benefit of much room for drying their clothes at the small caboose or cooking stove in the Barrack. For days together, also, the men were left without building materials, owing to the impossibility of landing them, or, what was worse, without the power of building what we had on hand in consequence of the violence of the winds.

Given the weather it seems remarkable that 12 courses comprising well over 1,000 stones were completed in the first six weeks. A further 13 courses would be laid by the end of the season.

Alan spends little time in his *Account* discussing course-by-course construction of the tower. Partly this is because the techniques were very similar to those used on the Bell Rock, already well documented by his father; and partly because a building project in a desolate location is extremely difficult to get started but, once under way and with meticulous planning, it gathers momentum as the workforce increases in experience and confidence. Starting from scratch in 1838, it took Alan three years to get the first few courses placed in Skerryvore's foundation pit; but by 1841 the tower was growing strongly and he felt free to go off on the occasional tangent.

The first of these concerns 'the illusive effects produced on the mind by the great waves which rolled past the Rock'. He was awed by their speed, noise, and size. Even when viewed from 30 feet up on the tower, they seemed about to wash right over the top and sweep all before them. It was only after a long stay on the rock and continual observation that he was able to 'mark the approach of their crested curling heads with composure' and risk a soaking by going to within a few feet of their expected reach. Even so, the swell produced by a strong north-westerly could produce some nasty surprises. His attempt to assess wave heights using a vertical rod inserted in the rock, graduated in large divisions that could be read at a distance, was a failure, but simple observation convinced him that large unbroken waves around Skerryvore typically measured 15 feet from trough to crest. Sailors in the steamer, tossed around and probably 'less accustomed to accurate measurement', tended to opt for 30 or even 40 feet.

He was fascinated by wildlife, including the most familiar of his mammalian companions:

> Amongst the many wonders of the 'great deep' which we witnessed at the Skerryvore, not the least is the agility and power displayed by the unshapely seal. I have often seen half a dozen of those animals round the Rock, playing on the surface or riding on the crests of curling waves, come so close as to permit us to see their eyes and head, and lead us to expect that they would be thrown high and dry at the foot of the Tower; when suddenly they performed a somersault within a few feet of the Rock, and diving into the flaky and wreathing foam, disappeared and as suddenly reappeared a hundred yards off, uttering a strange low cry, as we supposed, of satisfaction at having caught a fish. At such times the surf often drove among the crevices of the Rock a bleeding cod, from whose back a seal had taken a single moderate bite, leaving the rest to some less fastidious fisher.

Alan's next technical theme concerns the 'balance crane' which took over the task of laying stones when they reached the hollow part of the tower. The crane, a larger and stronger version of one used by his father on the Bell Rock, had been constructed in the Edinburgh workshop of a Mr James Dove. It was a crucial piece of heavy equipment for manoeuvring massive stones safely and precisely as the tower grew to dizzying heights above the sea. A cast-iron shaft, extendable as the building rose, was erected in the void. An iron frame, mounted on the shaft and able to rotate freely, carried two trussed arms fitted with gearing, chains, and rollers. One arm supported the stone, the other a counterpoise weight. The gearing was arranged so that any radial movement of the stone was matched by an equivalent movement of the counterpoise, keeping the crane balanced at all times. It was elegant, and it sounds simple enough, but great care was needed to keep the stone at the correct height as it moved inwards and swivelled round to precisely the right location. The task was entrusted to two men standing on small wooden platforms attached to the frame and rotating with it. In the final stages of the work they and their helpers were poised more than 130 feet above the sea on a tower which had shrunk to a diameter of just 16 feet. But there is never any mention of danger money.

On one occasion the balance crane was put to a severe test. A massive stone weighing nearly 2 tons had been raised by the exterior shear-poles and was being transferred to the crane. The tower diameter was still relatively large, so the weight of the stone had to be taken at the far end of the jib, exerting the maximum possible bending moment. Alan noted with alarm that the jib was deflecting 6–8 inches over its 14-foot length, yet the stone was manoeuvred and laid without incident and he subsequently detected not the slightest trace of damage. The balance crane was used 'with perfect confidence' for the rest of the season.

By the time rock operations finished on 17 August, 30,300 cubic feet of stone weighing 2,200 tons had been added to the tower. The stones were cut, and the courses laid, with such accuracy that the variation of diameter with height had kept within one-eighth of an inch of Alan's design hyperbola; and the top of the masonry, now at the 37th course, was within half an inch of the planned height. The first two 'apartments' in the void were

The balance crane, with its iron frame and two arms mounted on a central shaft. The left arm carries a counterpoise weight; the right arm, a roller and chain to take the stone.

complete and the third was well advanced. It was all extremely satisfying. They finished off the season's work by covering the balance crane with a strong tarpaulin and fixing a temporary lightning conductor from the top of the building down to the sea.

Alan finishes his *Account* for 1841 with a salute to the steamer *Skerryvore* and its master:

> The extent of work done during the season of 1841 at the Rock, must in a great measure be attributed to the advantage of steam attendance, without which numerous favourable opportunities of landing materials must necessarily have been lost, from the uncertainty which pervades all the movements of sailing craft. The number of lighters towed out and discharged at the Rock was 120; and it is remarkable that no accident of importance occurred, although many risks were run, from the breaking of warps while the craft lay moored to the landing quay during heavy seas.

But the successful season was to end in tragedy:

> I cannot omit in this place to record my sense of the services rendered to the works by the late Mr James Heddle, who commanded the steamer and who died from some consumptive disease soon after the close of the season's operations. Mr Heddle's health had been somewhat enfeebled towards the latter part of the autumn; and his excessive exertions and continued exposure during his arduous service, in some measure, I fear, hastened the crisis of his disease, which at length terminated suddenly by the rupture of an abscess in the lungs. Of his anxiety to forward the work, and his unwearied exertion

> in the discharge of his harassing duty at Skerryvore, which frequently allowed him less than twenty hours sleep in a week, I cannot speak too highly, as I consider his intrepidity and zeal to have been one of the most efficient causes of our success ever since the commencement of the works on the Rock in 1839. Mr Heddle possessed attainments superior to those generally found among persons in his walk of life and was in every respect a most estimable man.

So Alan, who had lost his trustworthy lieutenant Mr George Middlemiss to a heart attack at the end of 1840, now faced the near-certainty that Skerryvore had claimed another premature victim in 1841; not in a fall from the tower or by drowning at sea, but by the relentless pressure exerted on a human frame and spirit by a project so perilous that many considered it impossible. Alan, more than most, would have understood.

1842: Topping out

> On the 17th of April 1842, I made my first landing on the Skerryvore for the season, and found traces of very heavy seas having passed over the Rock during the preceding winter. Its surface was washed quite clean from all the scattered materials which were left lying on it at the end of the last season; and the building, to the height of 6 or 8 feet from the foundation, was covered with a thick coating of green sea-weed. The railway had suffered considerably from large stones having been thrown upon it ... Heavy sprays had been playing over the tower, in the upper uncovered apartment of which a great number of water-worn pebbles or boulders were found ... at a height of no less than 60 feet above high watermark ....

The winter had done its worst but the tower and barrack were unscathed. Alan's plans for the 1842 summer season included keeping about 20 seamen on the rock to help raise stones to the top of the tower, which still had 70 feet to go. They would need their own accommodation, so he decided to enclose the open platform immediately below the barrack kitchen with double planking and canvas. The resulting space, euphemistically referred to as an 'apartment', leaked like a sieve in bad weather but had its plus points in good:

> its inhabitants enjoyed more room, freer air, and more tolerable temperature, than any of their neighbours in the highest storey could obtain, owing to the greater number of persons in that part of the Barrack and its exposure to the heat of the cook's stove.

The rest of April and early May were marred by stiff gales which kept the men prisoners in the barrack and sluiced the windows of Alan's diminutive apartment, 55 feet above the sea, with sheets of water. It was 19 May before the 38th course – the first of the season – could be laid, but then came a welcome period of north-easterly winds, calm seas, and sunny weather. The seamen employed horizontal 'needles' projecting from the windows, and crab winches inside the tower, to raise stones to the top in stages. All went well until 9 July, a day of heavy seas, spring tides, and great anxiety. The stones of one of the highest courses, which had been delivered to the base of the tower, looked like disappearing into 'the insatiable deep', a disaster which could have delayed construction by a whole year. Fortunately the waves relented, subsequent landings were straightforward, and on 21 July:

> the last stones of the tower were safely landed on the Rock, under a salute from the steamer, as an expression, no doubt, of the satisfaction which the commander Mr Kerr and his crew naturally felt at having

> successfully brought out not fewer than 75 lighter loads, or about 1500 tons, of stone during the season, as well as in some measure of their joy at the prospect of a speedy and happy termination of our arduous labours.

On 25 July the tower was topped out with the last stone of the parapet, and the workforce removed the balance crane and its cast-iron pillar to make way for the lantern.

Alan, never one for elaborate ceremony, makes far less of the topping-out than his father would have done; but he does pause to reflect on the overall success of the building operations, due largely to the 'scrupulous accuracy of workmanship' and his insistence that the stones be dressed to within one eighth of an inch of his design specification. The elegant outline of the tower followed his favoured curve, the hyperbola, with extraordinary accuracy. Excellent mortar ensured that the masonry held fast and true, and even the ornamental cornice at the top of the building, which projected 3 feet out from the face of the wall, showed not the slightest sign of shifting.

The ornamentation, quite unnecessary from an engineering point of view, reflected Alan's keen artistic sense. He viewed the cornice as 'very bold and striking and quite in accordance with the simple and almost severe style of the pillar itself'. Even in a tower marooned 12 miles out to sea, visible only to the keepers and those whose lives it was designed to save, he felt it important to complement utility with beauty. The Institution of Civil Engineers would describe his tower as 'the finest combination of mass with elegance to be met with in architectural or engineering structures' and Robert Louis Stevenson, Alan's storyteller nephew, proclaimed Skerryvore 'the noblest of all extant deep-sea lights'.

After landing the final stone the steamer left Hynish for Greenock, towing two empty lighters to bring back the lantern. It returned to Skerryvore on 10 August. Two days later Robert Stevenson, just past his 70th birthday and on his final tour of inspection for the Northern Lighthouse Board, visited the rock to view his son's creation. Alan makes no comment on his father's reaction but it is easy to imagine a heady cocktail of emotions in the two men; one who had basked in 30 years of international acclaim for his Bell Rock masterpiece; the other, modest by nature, who had just completed a tower twice as massive and arguably more beautiful. A less highly charged meeting followed a week later when the Sheriff of Argyll entertained some gentlemen friends for breakfast at the base of the tower, climbed to the top, and 'minutely inspected every part of the work'.

By 14 September the lantern was assembled and glazed, ready to accept the light apparatus the following year:

> the glass was covered with a framework of timber to protect it from the sea-fowls which frequent in myriads the Rock and the Tower. The workmen were, on the same day, removed from the Rock, although with much difficulty, owing to the heavy surf which broke over the landing-place and rendered the embarkation more perilous than almost any I had before experienced at the Skerryvore.

Alan had survived a perilous first landing in 1838, and it proved almost as hard to depart in 1842. What he had built on that remote, storm-lashed rock in the meantime ranks as one of the great engineering achievements of the 19th century.

1843: Endgame

In January 1843, just four months after completing the Skerryvore tower, Alan was appointed Engineer to the Northern Lighthouse Board in place of his father. He was 36 years old. Although it was the ultimate professional accolade, its timing was a mixed blessing. The project on which he had laboured ceaselessly for 6 years was not quite finished and the light, a highly sophisticated lens system which he made his own speciality, had yet to be installed and commissioned. Just as he approached the magic moment when Skerryvore would shine out over the North Atlantic, he assumed overall responsibility for the Board's engineering activities – and his younger brother Thomas was put in charge of the remaining works at Hynish and on the rock.

In Chapter IX, 'Concluding operations and exhibition of the light', Alan describes the main tasks of 1843 – construction of the dock at Hynish, fitting out the interior of the tower and last but not least, installing the precious light.

A dock was needed at Hynish to shelter the vessel attending the lighthouse. It was a tough engineering challenge because of the rough weather and Atlantic swells that encumbered navigation around Hynish point, and Alan and Tom had given it a lot of

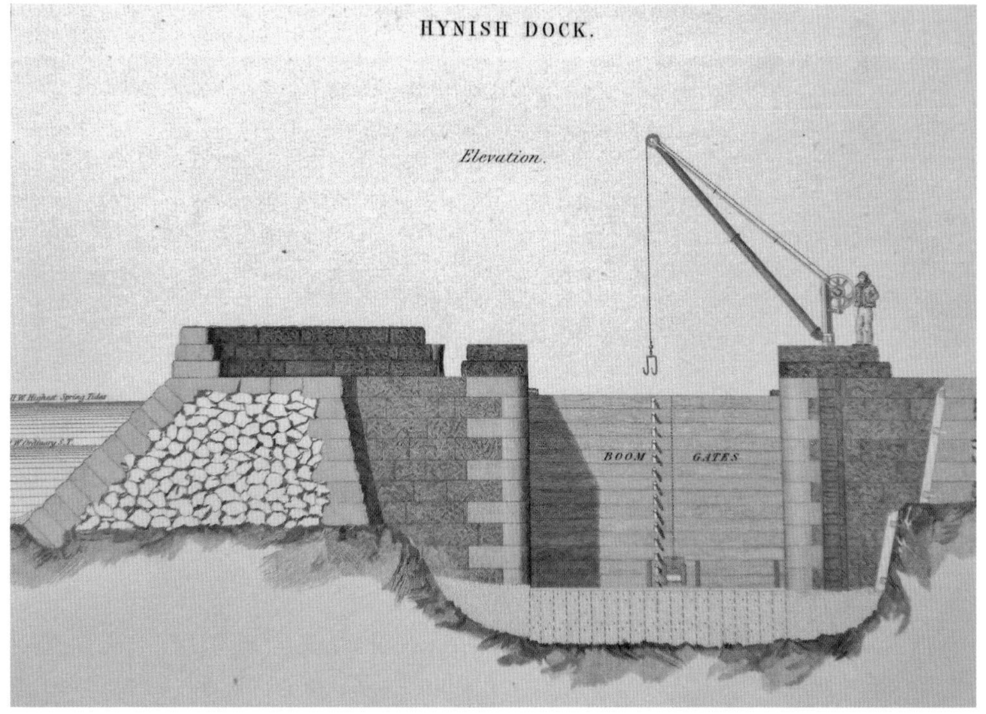

Construction details of Hynish dock.

thought. The vessel was not the valiant 150-ton *Skerryvore* steamer, soon to be sold, but a smaller tender of 35 tons named *Francis,* purchased at Deal in Kent. A basin 100 feet long and 50 feet wide with a depth of 12 feet at high water springs would provide shelter in all weathers and the chance to come and go safely in any sea condition that permitted landing on the rock.

The existing pier was first extended about 40 feet seaward of low water mark. Its vertical inner face, which had served as a quay for landing and shipping stones for the tower, provided one side of the new dock. Another, shorter, side and a gateway completed the basin, from which about 5,000 tons of rocky material had to be excavated to give the required water depth. The dock entrance could not be closed with conventional gates because of shifting sands, so timber booms, lowered into position one by one with a crane, were used instead. The design was unconventional and somewhat risky but, according to Alan, enjoyed 'the most complete success'.

Apart, that is, from one unforeseen problem. Heavy seas tended to build sand up against the outside of the boom gates and imprison the vessel within the dock:

> The only remedy for such an evil, was obviously to attempt some mode of artificial scouring; and for that purpose, it was proposed to divert several small streams which ran from Ben Hynish and the neighbouring hills through the grounds at Hynish, into one feeder, and pen them up in a pond, so as to afford the means of scouring the entrance to the basin from the incumbrance of the loose sand which might choke it.

So the slopes of Ben Hynish, 463 feet above the sea and overlooking the new dock, were to provide the water; and the brothers had to divert attention and resources to some unanticipated hydraulic engineering. Fortunately the offending material was Tiree's light shell sand which was 'easily acted on by currents of very feeble power' and they reckoned that a pond, or reservoir, containing 175,000 cubic feet of stream water, released by sluices through a conduit and into the dock, would provide one hour of effective scouring. Fortunately it worked a treat:

> The operation of scouring is performed at low water and is generally found quite sufficient for the purpose of clearing a passage down to the bare rock in a single tide. Nothing can be more satisfactory than to witness the effect of that process in opening the entrance to a basin apparently inaccessible; and but for such an arrangement, the dock must have remained permanently choked with sand and sea weed.

One episode in 1843 underlined the dock's vital contribution to the smooth running of the lighthouse. With the dock temporarily out of use, and no alternative available on Tiree, the tender was prevented from reaching the rock for seven anxious, seemingly interminable, weeks:

The dock at Hynish. The extended pier is visible at top right (photo: Tiree Images).

> The poor seamen who were living in the Barrack passed that time most drearily, for not only had their clothes been literally worn to rags, but they suffered the want of many things dearer to them than clothes, and amongst others of tobacco, the failure of the supply of which they had despondingly recorded in chalk on the walls of their prison house, with the date of the occurrence!

It was abundantly clear that the future success of the lighthouse, human and operational, depended on a reliable tender and a fully available dock at Hynish.

Tom, newly appointed Skerryvore's resident engineer, landed on the rock on 29 March 1843 and was delighted to find the lighthouse tower in excellent condition. It was only necessary to repoint the outside joints, a task carried out by men working on suspended cradles lowered precariously down the outside wall from the lantern gallery. He then turned his attention to the 'tedious operation' of fitting out the interior of the tower, including lining the walls of the apartments with wood, fixing flues for the fireplaces, and installing water tanks, oil tanks, and coal stores. Another facility, delightfully quaint by today's standards, was the provision of voicepipes for communication between the Light Room at the top and the various apartments. This was essential because a keeper on duty was forbidden to leave the Light Room on any pretext until relieved by a colleague:

> who is summoned by means of a bell placed inside his cot or sleeping berth, which is rung by means of a small piston, propelled by simply

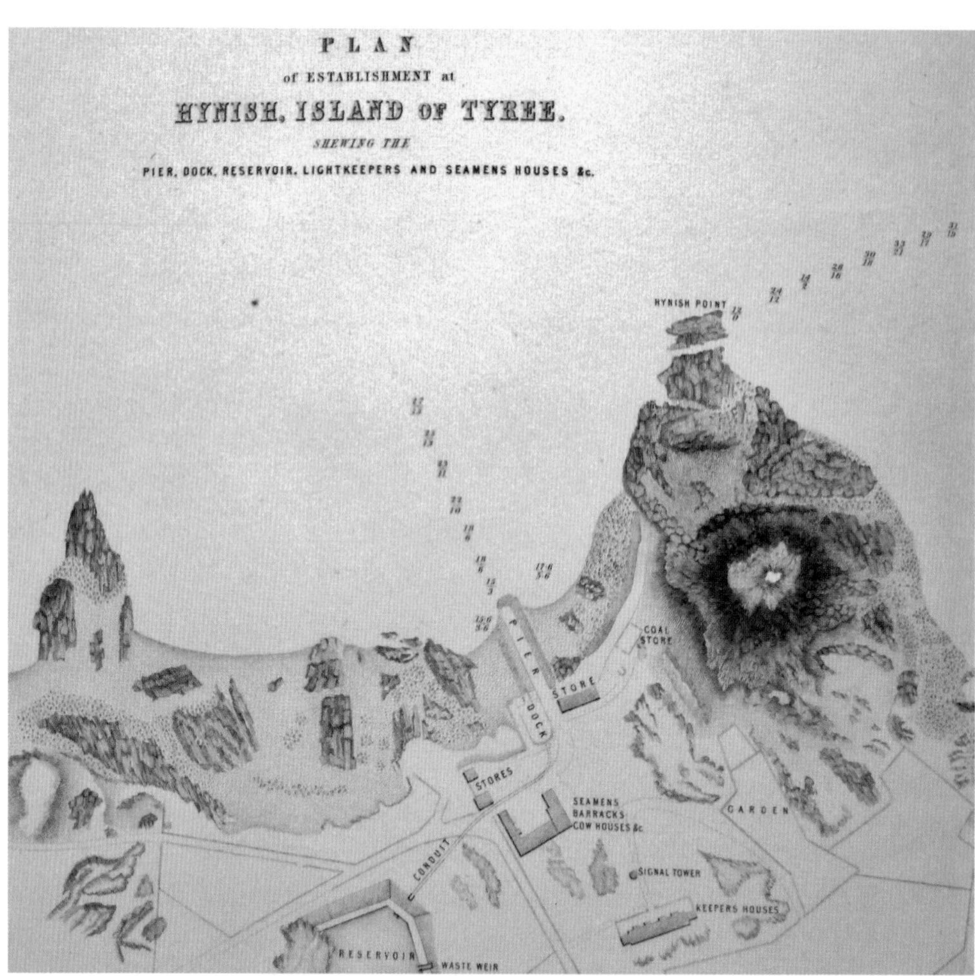

The completed establishment at Hynish including pier, dock, signal tower, keepers' houses, buildings and stores. Also shown are the reservoir and conduit used to scour out the dock.

> blowing into a mouth-piece in the Light-room. The keeper in bed answers this signal by a 'counter-blast' which rings another bell in the Light-room, and informs the keeper there that his signal has been heard and will be obeyed.

The entrance door to the lighthouse, 26 feet above the foundation pit, was reached by a vertiginous ladder of gun metal. Successive storeys were allocated to water-storage tanks; coals; a workshop; a provision store; a kitchen; two double apartments for the keepers; an apartment for visiting officers; an oil store; and finally, at the top, the Light Room and two fog bells. Oak trap ladders passed through hatches in each floor and were partitioned off from the apartments to prevent accidents and reduce cold draughts.

And now for the climax of Alan's *Account*, the northern light to crown his tower, the apple of his eye:

> The light is revolving, appearing in its brightest state once in every minute of time. It is elevated 150 feet above the sea, and is well seen as far as the curvature of the earth permits; it is also frequently seen as a brilliant light from the high land of Barra, a distance of 38 miles. The apparatus consists of eight annular lenses (of the first order, in the system of Augustin Fresnel), of 36.22 inches focal distance, revolving round a lamp with four concentric wicks, and producing a bright blaze when each lens passes between the lamp and the eye of a distant observer.

Skerryvore's lenses put it sea miles ahead of the reflectors installed by his father on the Bell Rock in 1811. Most of the lamp's light was concentrated by refraction through glass lenses (a dioptric system), as opposed to reflection by parabolic mirrors (a catoptric system), giving a roughly threefold increase in brightness. Alan was also encouraged by a long and fruitful collaboration with the French scientist Léonor Fresnel to add embellishments that captured light escaping above and below the eight main lenses. To appear 'brilliant' from 38 miles away on the Isle of Barra was certainly something to celebrate. Skerryvore pioneered a new era in Scottish lighthouse engineering, in sophisticated optics as well as tower construction.

A Fresnel lens uses less material and is more compact than a conventional lens because it is collapsed into a set of stepped, concentric, annular sections. Léonor Fresnel's elder brother Augustin had proposed six sizes of lens for use in lighthouses, and Skerryvore was equipped with the largest and most powerful type, referred to as first-order. They had a focal length, or distance, of 36.22 inches (920 millimetres), giving a complete assembly about 6 feet in diameter.

To capture the light escaping above, Alan followed a scheme invented by Augustin Fresnel. Eight smaller pyramidal lenses with a focal length of 19.68 inches (500 millimetres) were placed above the main lenses, inclined at 50 degrees to the horizon, together with eight plane mirrors inclined at 50 degrees in the opposite direction:

> By this arrangement, that part of the light from the lamp which would otherwise escape uselessly beyond the great lenses, upwards into the sky, being parallelized in its passage through the smaller lenses and falling on the mirrors, is finally projected forwards in horizontal beams, so as to aid the effect of the light. Those lenses and mirrors, however, instead of having their axes in the same vertical plane with the axes of the principal lenses, are inclined about 7° horizontally to the right hand, and by that deviation produce small premonitory blazes, which, blending with the beams of the larger lenses, tend in some measure to lengthen the duration of the impression on the eye.

Alan's second embellishment, modified from another Fresnel invention, was designed to collect the light that would otherwise escape downwards. He placed 'totally reflecting zones' below the main lenses, once again directing the light in the required direction – a technique explained more fully in the *Notes on the illumination of lighthouses* which form the second part of his book. The combination of reflecting and refracting elements produced a highly sophisticated optical system, sometimes referred to as catadioptric, which was manufactured for him in France. The associated machinery was constructed to Alan's 'entire satisfaction' by Mr John Milne in Edinburgh.

The proud moment for announcing Skerryvore's completion came in December 1843:

> The Commissioners of the Northern Lighthouses hereby give notice, that a Lighthouse has been erected on the Skerryvore Rock, which lies off the Island of Tyree, in the County of Argyll, the Light of which will be exhibited on the Night of the 1st of February 1844, and every Night thereafter, from sunset to sunrise.

The eight pencil beams of the revolving light blazed out across the North Atlantic as promised. The project, including all the works at Hynish and on the Ross of Mull, had cost a total of £90,268, to be offset by charging every British vessel which 'derived benefit from the light' a toll of one penny per ton, and every foreign vessel two pence per ton, on entering a British port. Nobody will ever know how many lives were saved, and wrecks avoided, in return for these modest payments.

A few months later the Commissioners had a marble tablet, inscribed in Latin, placed in the visiting officers' apartment near the top of the tower. Alan notes that the tablet:

> after acknowledging the hand of Almighty God in the success which attended the work, briefly sets forth the beneficent purposes for which the Lighthouse was erected and records the laying of the foundation-stone by his Grace the Duke of Argyll … .

He fails, however, to mention his own name inscribed at the bottom of the tablet. Instead, the remarkable *Account of the Skerryvore Lighthouse* ends with the words: 'I willingly close this most defective narrative of the work.'

Almost unbelievably, the designer and builder of 'the most graceful lighthouse in the world', the engineer who risked life, limb, and sanity to protect the lives of others, prefers to bow out with an apology.

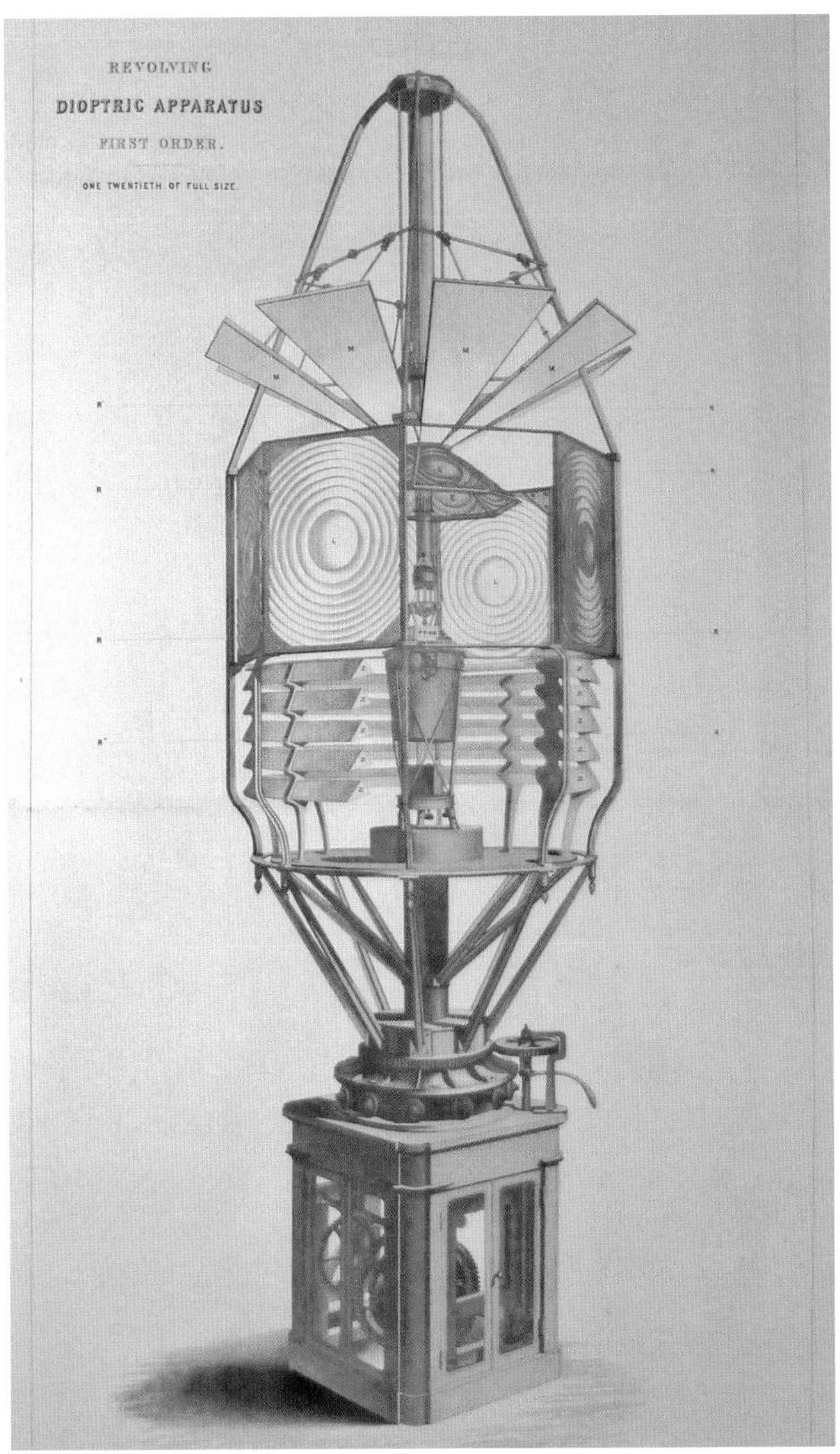

A revolving light of first order, showing eight main Fresnel lenses and the subsidiary lenses and mirrors designed to capture light that would otherwise escape above and below.

Alan Stevenson's magnificent tower (photo: Ian Cowe).

Engineer, scholar, artist

Alan was not quite finished with Skerryvore. As newly appointed Engineer to the Northern Lighthouse Board he was responsible, among a great many things, for defining the duties of keepers and advising on their salaries and conditions of employment. He knew Skerryvore's dangers and isolation only too well and recommended that the Commissioners pay the keepers a £10 premium to help compensate for 'almost complete banishment even from their families during several months of winter'. It was certainly a tough posting. The keepers had to move stores up and down the tower on their backs, negotiating the narrow staircases and hatches between the many floors. There was no bathroom or running water, they had to make do with buckets.

Yet in spite of the isolation Skerryvore became a favoured rock lighthouse among keepers who appreciated its internal space and the chance to take exercise on the rock in gentle weather. They had, of course, to put up with Hebridean storms; on many occasions, especially in winter, the lighthouse tender could neither land men and provisions on the rock nor rescue them. Unscheduled delays were particularly hard to endure and demanded great mental resilience, because contrary to popular belief it was often not loneliness that haunted the lighthouse keeper so much as the sharing of cramped personal space with colleagues. Skerryvore was not for the faint-hearted or over-imaginative, but for tougher members of the breed it could be wonderfully exhilarating.

On becoming Engineer Alan was immediately saddled with a host of other responsibilities, not least because the Board delayed appointing a new Clerk of Works. The major challenge was to construct more lighthouses. In fact he had already been working with Tom on the Little Ross light near Kirkcudbright, which was commissioned in the month of his appointment (and destined to witness the 1960 murder of an assistant keeper by a colleague). By the time Alan retired in 1853 he had built thirteen lighthouses of which eight were classed as 'major lights'. Apart from Skerryvore the best known is Ardnamurchan, with its Egyptian-style features, lit in 1849. Perched on the westernmost tip of the Scottish mainland, its proximity to Mull meant that Alan had no difficulty in deciding on stone for the tower – it had to be his favourite pink granite. The other newcomers were scattered around the coast of Scotland, from Noss Head in the far north-east to Sanda in the south-west. When Alan retired the Board's portfolio was approaching 40 lighthouses but only two, Bell Rock and Skerryvore, were on isolated rocks and his own creation was the tallest of all by a considerable margin.

Alan also took over the annual tours of inspection of Scottish lights that had been a major feature of his father's tenure as Engineer. But there were far more lighthouses than Robert had started with, and the organisation and exertion involved were daunting. On one windless occasion Alan travelled over 30 miles in an open rowing boat to inspect the Barra Head light; on another he landed at Cape Wrath and had to struggle 200 feet up a

Skerryvore in gentle weather (Photo: Ian Cowe).

Three of Alan Stevenson's later lighthouses: Ardnamurchan, 1849 (Wikipedia); Noss Head, 1849 (NLB); and Sanda 1850 (NLB).

vertiginous, pathless cliff to inspect the lighthouse. Long, damp and dangerous voyages around the coast were emphatically not what the doctor ordered for a man whose health had always been suspect and he began to suffer bouts of chronic tiredness, rheumatism, fevers, and aching in the limbs. But he soldiered on with all the duties that came with the job, including the committee work and political lobbying that so ill-suited his temperament. What Alan excelled at was the single-minded attention to a project that fired his imagination, for which he had infinite patience.

There is, perhaps, a wider issue. One of the great engineers of the Victorian era had just risked the best years of his life designing and building a wonderful lighthouse in an almost impossible location. Like the mountaineer who has climbed Everest and cannot find a better peak to conquer, or the athlete who has devoted four years to winning Olympic gold and struggles to refocus, Alan must surely have felt some measure of anti-climax. In such situations it is common for the effects of stress, ignored or suppressed over a long period, to rise to the surface. He would probably have been wise to take a year off after Skerryvore but it was not on the cards with the Commissioners, nor was it in line with prevailing Scottish or family attitudes towards Christian duty. Whatever the reasons, his health was already on a downward path.

The special talents that Alan brought to Skerryvore are well illustrated in his *Account* which, in spite of his protestations to the contrary, is one of the finest engineering publications of the 19th century. As we have seen, he thrived on problems with an intellectual content and investigated them fully – examples being his mini-treatises on the mining and fine dressing of granite, the redesign of the temporary barrack after the catastrophic failure of 1838 and, above all, his masterly chapter 'On the Construction of Lighthouse Towers'. These are not the fruits of a butterfly mind. It must have been hard to swap Skerryvore for a role that involved endless chopping and changing between different tasks, heavy but fragmentary responsibilities, and interruptions due to crises in far-flung places that were not of his making. He hints at all this in the preface to his *Account* when he apologises for its supposed 'defects':

> My chief plea in defence is, that the preparation of this Account of the Skerryvore Lighthouse, and the Notes on the Illumination of Lighthouses which follow it, was not chosen or assumed by me, but was a task imposed by the express desire of the Lighthouse Board, to whose enlightened and liberal views the Mariner owes the erection of the Lighthouse itself. My labours were also continually interrupted by the urgent calls of my official duties; and, on several occasions, I was forced to dismiss unfinished chapters from my mind for a period of several months – circumstances which, I hope, will in some measure account for the desultory character of the performance, the disproportion of some of its parts, and more specially for repetitions and perhaps omissions which would otherwise have been quite unpardonable.

In fact his publication appeared 4 years after Skerryvore was lit – comparing rather favourably with the 13 years it had taken his father to produce the *Account of the Bell Rock Lighthouse.* But then Alan was far happier with a pen, and could write with confidence and speed when time permitted.

His *Account of the Skerryvore Lighthouse* carries the subtitle *with notes on the illumination of lighthouses*, suggesting a modest addition to the main narrative. But this is far from the case – the 'notes' form a major treatise which, at 169 pages, is as long as his description of Skerryvore itself. It covers a crucial aspect of lighthouse engineering that Alan made particularly his own, and is a striking example of his scholarship – the ability to study and digest state-of-the-art optical theory and apply it to capturing and focusing as much lamp light as possible. Packed with design details and supported by extensive geometry and trigonometry, it is a tour-de-force that proved highly influential for a hundred years. Just as impressively, Alan put theory into practice when, as Engineer to the Board, he began the process of re-equipping all the Scottish lights with dioptric (lens) systems in place of the catoptric (reflecting mirror) systems developed by Thomas Smith and his own father.

Alan's fascination with lenses stemmed from a visit to France in 1824 when he met the famous Fresnel brothers, world leaders in lens design for lighthouse illumination. Tragically Augustin Fresnel, who had pioneered a dioptric system for the Tour de Cordouan light in 1822 and was elected a member of the Royal Society of London soon afterwards, died of tuberculosis in 1827, but when Alan was sent back to France by the Commissioners in 1834 he formed a lasting friendship with the younger Léonor. Alan was encouraged to visit lighthouses around the French coast and sent no less than 17 diplomatic letters back to his sceptical father reporting what he had seen. He never missed an opportunity to praise the Fresnels' work and the generous hospitality of a like-minded colleague across the Channel, so recently the dividing line between two nations at war:

> I was much encouraged by the singular liberality of Mr Léonor Fresnel, to whose friendship (as I have often, with much pleasure, acknowledged) I owe all that I know of dioptric Lighthouses. He not only freely communicated to me the method pursued by his distinguished brother Augustin Fresnel, in determining the forms of the zones of the small apparatus, introduced by him into the Harbour Lights of France, and his own mode of rigorously solving some of the preliminary questions involved in the computations; but put me in possession of various important suggestions, which substantially embrace the whole subject.

Indeed, so firm did their friendship become that he happily escorted Léonor and his wife on a tour of the Scottish lights in 1837, reporting from Thurso in the far north that 'Fresnel is in high spirits … he and Madame sung a scene of an opera after breakfast in capital style'.

Alan's treatise on illumination is far more than a technical triumph. He is incapable of ignoring the wider aspects of any subject he tackles, and loves to inject historical, literary, and artistic allusions. He starts off with a few delightful pages on the lighthouses of antiquity, especially the Pharos of Alexandria, with copious quotations from Greek and Latin sources, and moves on to famous European examples including the Tour de Cordouan, the Eddystone and finally the Bell Rock. He discusses, in considerable detail, the historic progression from illumination by wood and coal fires to candles and oil lamps; the intensity and form of the flames produced; and the special features of catoptric and dioptric systems. As always there is a social context and he is especially concerned that, as the number of lights around a coastline increases:

> their very number itself produces a new evil, in the difficulty of distinguishing the lights from each other. As the object of a light is to make known to the benighted mariner the land he has made, with

252 NOTES ON THE ILLUMINATION OF LIGHTHOUSES.

I come next to the *second* case, which concerns the calcula-

Let C be the centre of curvature (see fig. 57)

$\alpha = ACb$ the angle of emergence

$\eta = B'bC$ the second angle of refraction

$\epsilon = BbB'$ the first angle of refraction

$i = B'FA$ the first angle of incidence

$i' = BFA$

$e = b'Bb$

$AB \doteq r$

$AB' = r'$

$Bb = t''$ the thickness of the lens at the edge

$AF = \phi$ the focal distance.

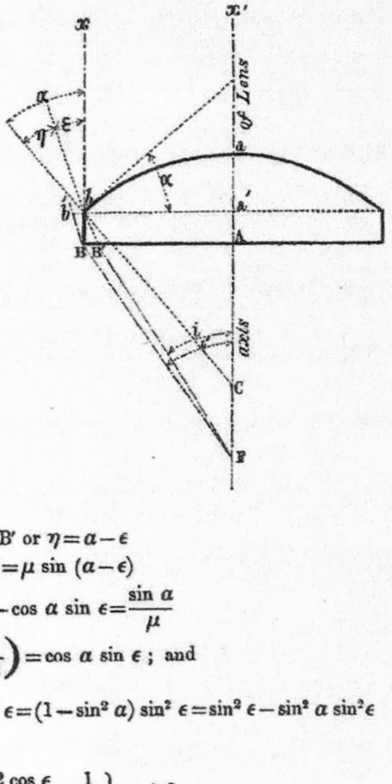

Then $\tan i' = \frac{r}{\phi}$; $\sin e = \frac{\sin i'}{\mu}$

whence $bb' = t'' \tan e$ becomes known.

Now, since $BB' = bb'$ nearly, $AB' = AB - bb'$

or $r' = r - t'' \tan e$.

From this is obtained the angle of incidence i, and the first angle of refraction ϵ;

for $\tan i = \frac{r'}{\phi}$ and $\sin \epsilon = \frac{\sin i}{\mu}$

Fig. 57.

Next $B'bC = BbC - BbB'$ or $\eta = \alpha - \epsilon$

and $\sin \alpha = \mu \sin \eta = \mu \sin(\alpha - \epsilon)$

from which, $\sin \alpha \cos \epsilon - \cos \alpha \sin \epsilon = \frac{\sin \alpha}{\mu}$

whence $\sin \alpha \left(\cos \epsilon - \frac{1}{\mu}\right) = \cos \alpha \sin \epsilon$; and

$$\sin^2 \alpha \left(\cos^2 \epsilon - \frac{2 \cos \epsilon}{\mu} + \frac{1}{\mu^2}\right) = \cos^2 \alpha \sin^2 \epsilon = (1 - \sin^2 \alpha) \sin^2 \epsilon = \sin^2 \epsilon - \sin^2 \alpha \sin^2 \epsilon$$

Then transposing we have

$$\sin^2 \alpha \left\{ (\cos^2 \epsilon + \sin^2 \epsilon) - \frac{2 \cos \epsilon}{\mu} + \frac{1}{\mu^2} \right\} = \sin^2 \epsilon$$

and because $(\cos^2 \epsilon + \sin^2 \epsilon) = 1$ we have, by dividing,

$$\sin^2 \alpha = \frac{\sin^2 \epsilon}{\left\{1 - \frac{2 \cos \epsilon}{\mu} + \frac{1}{\mu^2}\right\}} = \frac{\mu^2 \sin^2 \epsilon}{\mu^2 - 2\mu \cos \epsilon + 1}$$

$$\text{and } \sin \alpha = \frac{\mu \sin \epsilon}{\sqrt{1 - 2\mu \cos \epsilon + \mu^2}}$$

Next, since $bC = \frac{a'b}{\sin ACb} = \frac{r}{\sin \alpha}$, putting $Cb = \rho''$, and substituting

$$\text{we have } \rho'' = \frac{r}{\mu \sin \epsilon} \sqrt{\mu^2 - 2\mu \cos \epsilon + 1}$$

and, taking for the radius of curvature, the mean of ρ' and ρ'' the values calculated for the central and marginal rays, we have finally $\rho = \frac{\rho' + \rho''}{2}$

This page from Alan's Notes on the Illumination of Lighthouses gives some idea of the mathematical sophistication that he and the Fresnels brought to their work on lens design.

> as much certainty as the sight of a hill or tower would shew him his position during the day, it becomes an object of the first importance to impress upon each light a distinctive character, which shall effectually prevent the possibility of its being mistaken for any other.

All of which leads to a discussion of flashing, revolving, and coloured lights, and the whole complex business of efficient focusing with mirrors and lenses.

Alan is by now about halfway through his treatise and much of the rest is detailed and mathematical. He is, of course, working towards a description of the catadioptric (combined lens and mirror) apparatus he had designed for Skerryvore in conjunction with Léonor Fresnel and his master lens maker François Soleil:

> who at once boldly undertook to furnish for the Skerry vor Lighthouse the first catadioptric apparatus ever constructed on so magnificent a scale. On the 23rd December 1843, M. Fresnel announced, in a letter to me, the complete success which had attended a trial of the apparatus at the Royal Observatory at Paris. I know no work of art more beautiful or creditable to the boldness, ardour, intelligence, and zeal of the artist.

The Skerryvore apparatus was easily the most powerful and sophisticated light ever installed in a Scottish lighthouse and Alan leaves us in no doubt of his view that the finest engineering, especially when performed 'in the service of man', is an art as well as a science.

Alan was not a painter but he was certainly an artist in the wider sense. This side of his temperament was expressed in many ways, including the embellishment of lighthouse towers with architectural features that added beauty to utility. He seems to have designed cast-iron birds, crocodiles and a rustic balustrade with animal feet for his father's lighthouse at Girdle Ness (1833). He certainly spent a lot of time and thought on Skerryvore's cornice and added delightful iron hand grips in the form of sea serpents. Later, he decorated the Ardnamurchan tower in Egyptian style to reflect his admiration for the ancient Pharos of Alexandria.

Another passion – his love of the natural world – shines out from his writing. Early in the *Account* he notes that the surface of Skerryvore's rock was rent by deep gullies or fissures, one of which terminated in a hollow submarine chamber capable of ejecting a 20-foot water spout:

> resembling in appearance the Geyser of Iceland, and accompanied by a loud sound like the snorting of some sea monster. The effect of this marine jet d'eau was at times extremely beautiful, the water being so much broken as to form a snow-white and opaque pillar, surrounded by a fine vapour, in which, during sunshine, beautiful rainbows were observed. But its beauties by no means reconciled us

> to the inconvenience and discomfort it occasioned, by drenching us whenever our work carried us near it.

On another occasion he describes how:

> One beautiful morning, during our stay of six days at the Rock, we had a visit from a shoal of small fish, whose novel appearance made me take them for a fleet of some species of Nautilus. Those animals came in such numbers, that the pale blue silky membranes or sails, which wafted them before a gentle breeze over the glassy surface of the ocean, literally covered the water as far as we could see.

And, in spite of days spent in exhausting labour, he feels able to assure us that:

> life on the Skerryvore Rock was by no means destitute of its peculiar pleasures. The grandeur of the ocean's rage, the deep murmur of the waves, the hoarse cry of the seabirds, which wheeled continually over us, especially at our meals, the low moaning of the wind, or the gorgeous brightness of a glassy sea and a cloudless sky, and the solemn stillness of a deep blue vault, studded with stars or cheered by the splendours of the full moon, were the phases of external things that often arrested our thoughts in a situation where, with all the bustle that sometimes prevailed, there was necessarily so much time for reflection.

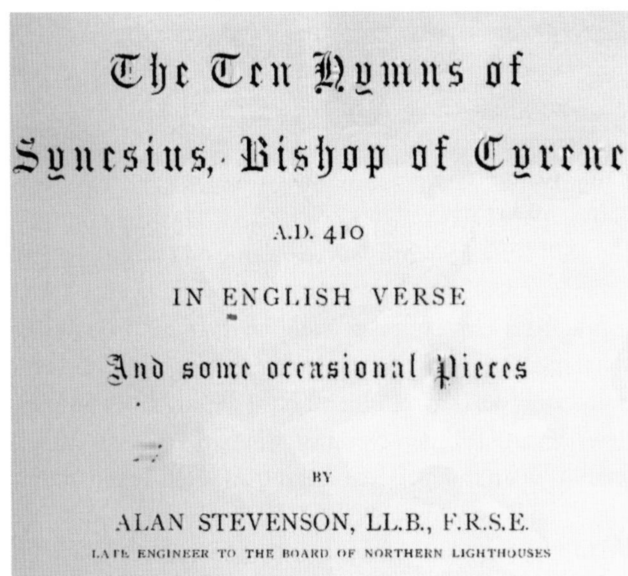

The Ten Hymns of
Synesius, Bishop of Cyrene
A.D. 410
IN ENGLISH VERSE
And some occasional Pieces
BY
ALAN STEVENSON, LL.B., F.R.S.E.
LATE ENGINEER TO THE BOARD OF NORTHERN LIGHTHOUSES

Alan's translation of the Hymns of Synesius was printed for private circulation by his family in 1865, the year of his death.

It was hardly necessary for Alan to decorate the *Account* with such passages because, as he himself noted, it was a 'task imposed by the express desire of the Lighthouse Board' whose priority was an accurate record of Skerryvore's design, construction, and costs. But their Engineer could not desist. He spent countless hours on the rock between 1838 and 1842 and needed to acknowledge precious moments spent in wonder at the world around him.

Alan was completely at home with Ancient Greek and Latin, and highly proficient in French, Italian, and Spanish. In the evenings, or when imprisoned by bad weather in his tiny apartment in the temporary barrack, he would indulge a love of literature and poetry by reading from his miniature library of favourites, including *Don Quixote,* Aristophanes and Dante. He even translated the fifth-century *Hymns of Synesius* from the Greek and sent them to an old university friend.

Alan had a great sense of humour and in his youth could see a joke in almost any situation. Although he parodied many of the visitors to the family's Edinburgh home, one of them, in particular, impressed him. Sir Walter Scott, 36 years his senior, knew his father well and sang the praises of the Bell Rock lighthouse in his poem *Pharos Loquitur*. Although Scott died before Alan started on Skerryvore, he had visited the rock with Robert in 1814 and found it a formidable site for a lighthouse, 'the Bell Rock and Eddystone a joke to it'. The great novelist would have been astonished to see the finished product.

Sir Walter Scott (1771–1832) and William Wordsworth (1770–1850) (Wikipedia).

Alan also struck up a friendship with William Wordsworth, relying on correspondence rather than face-to-face contact. The engineer and the romantic poet developed a strong mutual respect, springing on one side from a love of poetry, and on the other from an appreciation of the enormous personal challenges of lighthouse engineering. Alan declared that 'many were the moments in my solitude, during which I have felt my commonplace labours ennobled by the poet's views of duty and perseverance', whereas Wordsworth valued above all the heartfelt appreciation of an independent mind from outside the literary establishment. Their point of contact had been a mutual friend, James Wilson, who sometimes visited Wordsworth and told him about Alan's adventures on Skerryvore. This led to correspondence and Wordsworth even sent Alan some mementoes including a lock of hair and a few laurel leaves that ended up on the wall of his storm-lashed apartment. When the *Account* was published in 1848 Alan sent a copy to the 77-year-old poet and received, in return, a cherished letter:

Rydal Mount
Ambleside
May 20th 1848

My dear Sir

Accept my grateful thanks for the valuable Present of your Account of the Skerry Lighthouse &c from the perusal of which I promise myself confidently both instruction and pleasure, in no small degree. Every one who thinks and feels must take a lively interest in your lonely situation and most important employment. This be assured I do eminently, and with sincere good wishes for your health and well-being.

I remain, my dear Sir, faithfully your much obliged

Wm Wordsworth

So what were the mainsprings of Alan's decision to become a lighthouse engineer and forgo a more comfortable life immersed in poetry, literature and the arts? On the face of it a man with his intelligence, sensitivity, and indifferent health would have been better suited to life as a university don in his home city of Edinburgh or, given his strong religious convictions, a country clergyman. Yet he subjected himself to a profession which in the middle of the 19th century was almost guaranteed to place intolerable demands on his constitution.

The conundrum brings to mind the 'nature versus nurture' debate, one of the oldest battlegrounds in psychology. What are the relative contributions of genetic inheritance and environmental factors to personal development; is the newborn child a blank slate, or are its characteristics largely inborn; what effects do family and societal pressures have on a person's life choices? Today most experts agree that nature and nurture both play key roles, but the extent of each remains a hot topic. No doubt characteristics such as eye and hair colour are biologically determined; but what about life expectancy, intelligence, ambition, and human empathy? As we move from the physical to the emotional, the waters become increasingly murky.

When a couple such as Robert and Jean Stevenson have multiple children, they are often remarkably different owing to the random mix of genes each receives from the parents. The three surviving sons of Robert and Jean who went on to become in turn Engineer to the Northern Lighthouse Board were good examples, and of the three it seems clear that baby Alan had been dealt the most challenging hand. Some aces and court cards, certainly; but also, given his subsequent choice of career, a few from nearer the bottom of the pack.

If the genetic lottery had treated the three boys unequally, the environment in which they grew up was largely shared. All lived in a large and fashionable Edinburgh home with a fine garden, all were watched over by two forceful parents who loved them and tried to control them. Their father was tough, conservative, and successful, with strong views about their education; their mother, who had lost four babies, was highly anxious about the health of the survivors – and especially Alan. Both parents were conventionally religious in the Scottish protestant tradition that seemed, to some, unnecessarily joyless and dosed with guilt. As adulthood approached the boys were left in no doubt that hard work and a firm faith were the keys to success, even though they had plenty of mischievous fun together in the meantime.

All this produced the kaleidoscopic mix of nature and nature that went into the formation of Alan. He had, in addition, to cope with the bane of many an eldest son, the unwavering assumption that he would follow his father into the family business. Robert exerted relentless pressure on Alan over many years and eventually wrote him a formal letter demanding a decision. The reply was couched in cheeky language that probably irritated, but it gave Robert the desired result:

> Dear Father,
> I take this opportunity of answering your letter in which you stated a desire that I would apply myself to some business and although I must confess I had a liking for the profession of a soldier, on receipt of your letter I determined to overcome this foolish wish and am happy to say I have succeeded. On further consideration I found in myself a strong desire for literary glory and I picked upon an advocate but there was a want of interest. It was the same way with a clergyman and as I am by no means fond of shop-keeping I determined upon an engineer, especially that all with whom I have spoken on the subject recommend it and as you yourself seemed to point it out as the most fit situation in life I could desire. I only doubt that my talents do not lie that way, but in hopes that my choice will meet with your full approbation,
>
> I remain, my dear Father,
>
> Your ever affectionate and grateful son,
>
> Alan Stevenson.

Having made up his mind – or, some would say, capitulated – Alan never looked back, and committed a generous slice of heart and soul to the job.

The health issue, which dogged him all his life, was curious. Often sickly as a child, with long periods confined to bed, he seems to have done rather better during the later school and university years. His father tried to toughen him up by exposing him to all sorts of physical challenges, and was always suspicious of his love of poetry and literature.

Alan countered Robert's temperamental inability to understand him with a mixture of intelligent humouring and escape. At home in Edinburgh under the parental gaze he tended, even as an adult, to succumb to illness; away from home, especially on the travels to France, Russia, Sweden, and Holland that he enjoyed so much, his spirit soared and his health responded. The Skerryvore years were a very mixed blessing: escape of a sort for a few months each summer, but into an enormously demanding physical environment; winters back in Edinburgh threading his way through a minefield of paperwork and personal relationships. As we have seen, by the time he became Engineer to the Board, his health, though not yet broken, was clearly heading south.

Fortunately one of his excursions from Edinburgh as Clerk of Works to the Northern Lighthouse Board was to transform and charm his personal life. At the age of 26 he visited Anglesey with his brother David to conduct a survey. He met 21-year-old Margaret Scott Jones at a ball and, unsurprisingly for a man who surely needed the love of a young woman, he fell for her. Margaret, who lived with a strict Victorian father, two sisters, and three brothers at Llynon Hall surrounded by beautiful countryside, must have felt the same because she waited 11 years for him. By the time he returned to marry her in September 1844, he was Engineer to the Board with an astonishing set of achievements to his name, including, above all, Skerryvore.

The 11-year wait was also hard for Alan. As his professional responsibilities multiplied there must have been many occasions when he longed for the love of his life, but he accepted the view of his father that work comes first and was determined to see Skerryvore through

To the 'Ringing Stone' at Balaphetrish.
Tyree.

MYSTERIOUS Stone! rude, shapeless as thou art,
Thou seem'st unconscious of the ocean's rage.
Or winter-tempests that for many an age
Have howl'd around thee; say, hast thou a heart
Deep prison'd in thy mass, that feels the smart
Of others' woes—woes of the gentler kind,
Which spring up easily in woman's mind!
For, touch'd by maiden's hand, with gentle art,
Thou givest sighs, that tremble on the breeze,
Which sweeps around the western Hebrides,
Such as Andromeda, from ocean's cave,
Might breathe responsive to some sorrowing maid
Whom slighted vows or dear hopes long delay'd,
Have driven to seek, near thee, a lonely ocean-grave.

HYNISH, *May* 13, 1842.

Alan's poem to the famous 'Ringing Stone', a large boulder balanced on other rocks beside Tiree's northwest coast.

The Stevensons in 1860. Alan, aged 53 and by now head of the family, sits in the centre with Margaret and their son Robert, aged about 13, behind him. On the right is David, flanked by a nanny, daughters and wife Elizabeth. Centre left, at the back, Thomas stands with wife Maggie and their son Robert Louis, aged about 9. (© Leslie and Paxton, 1999).

before finally committing to Margaret. In the meantime they would correspond, he would write her poetry, they would live in hope. On one occasion in 1842, imprisoned by Atlantic storms for a fortnight in his minuscule apartment on Skerryvore, he dwelt morosely on the dreadful possibility that he might lose her. His poem 'To the "Ringing Stone" at Balaphetrish, Tyree' is a poignant reminder of his vulnerability in matters of the heart.

Marriage was the one great escape Alan needed, this time into family life with Margaret and away from his father's insistent gaze. Their union produced three daughters and a son, Robert – who would become an art critic and teacher – and lasted 21 years until Alan's death. The newlyweds moved to Windsor Street and then to the delightful Regent Terrace in Edinburgh, where their daughters enjoyed a beautiful garden. Alan continued to write poetry for his wife and children expressing the love he had for them, and we have to believe that his decline was at least partly softened by the love they showed for him.

Increasingly disabled by a progressive paralysis, Alan resigned as Engineer to the Northern Lighthouse Board in 1853 at the age of 46, and was succeeded by his brothers David and Thomas. A request to the Treasury for a pension proved fruitless and financial help received from the Commissioners was eroded by equally fruitless medical treatment. The family moved to a smaller house in Portobello, Edinburgh, and later to the country village of St Cyrus, some 20 miles north of Arbroath. Alan continued to write and showed extraordinary courage in the face of suffering. When he died in Edinburgh two days before Christmas 1865 the death certificate's medical insight was summarised by a terse 'General paralysis – 8 years'.

Alan remained deeply religious throughout his adult life, with a faith that was often burdened by feelings of guilt and inadequacy. Brought up in the strict traditions of Scottish protestantism, he became attracted to Roman Catholicism during his visits to France, but eventually settled for the established church back home. In the *Account* of Skerryvore he often gives thanks to his God for delivery from the extraordinary dangers faced at sea and on the rock. Coming from some people such words might be viewed as mere 19th-century convention, but it is impossible to doubt Alan's deep sincerity. As his health deteriorated, it seems he came to regard his suffering, at least in part, as divine retribution for sins committed, the 'wrath of God' so often mentioned in the Christian Bible.

As early as 1846, just two years after Skerryvore was lit, Alan was finding time in an enormously busy work schedule to worry about his imagined misdeeds. During the construction of the lighthouse he had appealed to his workforce to turn out on Sundays, defying social and religious convention very much as his father had done on the Bell Rock. Absolutely nothing moved in Tiree on a Sunday and it is easy to imagine raised eyebrows and village mutterings about a man from Edinburgh who presumed to violate the Sabbath. This troubled Alan so much that he sent a letter to a former work colleague who had taken a post as Assistant Keeper on the Pentland Skerries lighthouse:

> Dear Sir,
>
> I feel it to be my duty to address a few words to you on the subject of the greatest moment, which I habitually disregarded while you were engaged some years ago at the Skerryvore Lighthouse works under my direction. The difficulty of the work in which we were engaged, and the uncertainty of the weather, I at that time considered as a sufficient plea for violating the sanctity of the Lord's Day; but I have now seriously to declare my deliberate and sober conviction that, although there be works of necessity and mercy which ought to be done on the Day of Rest (according to the precept and example of our Blessed Lord himself), it may well be doubted whether almost any of the operations at the Skerryvore were of the kind to be so exempted … The Christian Sabbath was openly profaned, almost without the shadow of pretext.

The reply, assuming there was one, has not survived. It is to be hoped that in contrast to the tenets of an uncompromising religion it contained at least a modicum of human compassion. And even if not, we may be sure that Alan's later years were made more bearable by the love and respect of his family and friends. When he died the Commissioners of the Northern Lighthouse Board recorded their 'deep and abiding regret for the loss of a man

… whose genuine piety, kind heart and high intellect made him beloved'. Whatever Alan thought of himself, he was treasured by those around him.

It is given to very few people to match the breadth and quality of Alan Stevenson's achievements. But greatness must surely be measured by the effort made and price paid as well as the results. In all three departments his contributions were exceptional; and as we turn the final pages of the *Account of the Skerryvore Lighthouse* the only possible conclusion, even at this remove of space and time, is that we have been in the company of a great man.

Epilogue

As evening approaches the northern light of Skerryvore flashes out once more – as it did without fail for 110 years until, one night in March 1954, a disastrous fire broke out on the seventh floor, spread rapidly, and forced the keepers out onto the rock. Nobody died, but the damage was distressing – the light and interior largely destroyed, masonry cracked. Yet the engineer's structure survived not only the storms it was designed for, but also a cruel attack from within.

The restoration, awkward and lengthy, involved complete renovation of the interior and conversion of the light to electricity. Five years later the lighthouse resumed duty, flashed white across the North Atlantic, and regained its dignity.

As evening approaches Alan Stevenson embarks on a final voyage, an act of his creative imagination. Clutching the *Account*, Margaret at his side, he sets out once more for Tiree and treads the path to Hynish. The dock is in good order – Tom must have scoured it out for them; the tender is in steam – Mr James Heddle must have known they were coming. As the tide fills they round the point and head out across the Passage of Tiree in which 'no vessel can live'. Can that be Mr George Middlemiss standing in the wave-tossed bow, come to check the temporary barrack? The steamer ploughs on, twin engines throbbing, course set WSW into the fading amber of a Hebridean sky.

A pencil beam pierces the gathering darkness and, unhurried but insistent, guides the mariners towards the deep black gneiss of Skerryvore. Rocky outliers pose no threat, Boinshly and Bo-rhua no terror. Hand in hand but free of charts and work schedules, Alan Stevenson is finally a man unhindered. And as the boat nears home he understands, and accepts to the core of his being, that in spite of all the setbacks this was a work well done, a graceful structure strongly built, a signal service rendered *in salutem omnium*.

Appendix: Weights and measures

Alan Stevenson used imperial units in his *Account of the Skerryvore Lighthouse*. Metric equivalents are as follows:

LENGTH

- 1 inch = 25.4 millimetres
- 1 foot = 0.305 metre
- 1 mile = 1.61 kilometres

AREA AND VOLUME

- 1 square foot = 0.0929 square metre
- 1 cubic foot = 0.0283 cubic metre

MASS

- 1 pound = 0.454 kilogram
- 1 imperial ton = 1.016 tonnes

SKERRYVORE LIGHTHOUSE

- Total height of tower: 156 feet = 47.6 metres
- Height of light above high water spring tides: c. 150 feet = 45.8 metres
- Height of masonry: 138 feet = 42.1 metres
- Height of shaft (hyperbolic profile): 120 feet = 36.6 metres
- Diameter at top of shaft: 16 feet = 4.9 metres
- Diameter at base of shaft: 42 feet = 12.8 metres
- Volume of masonry: 58,580 cubic feet = 1,658 cubic metres
- Weight of masonry: c. 4,300 imperial tons = 4,369 tonnes

Index

Note: figures in bold type indicate an illustration.

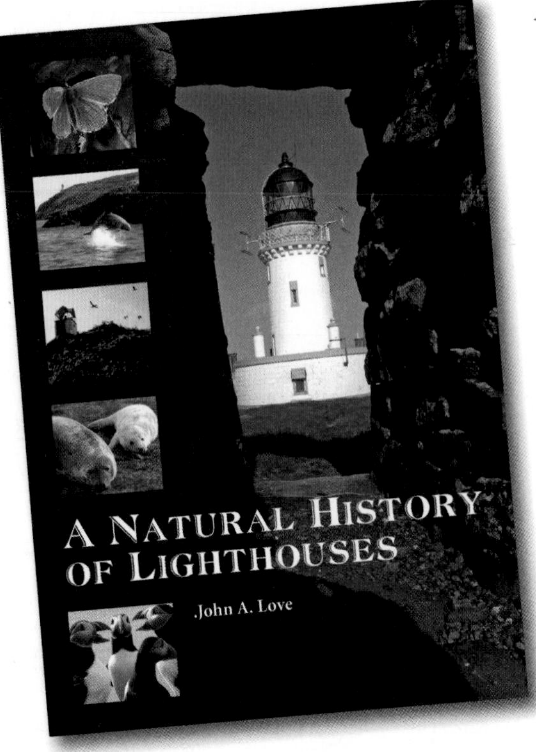

A Natural History of Lighthouses

- A history of lighthouses and the Stevenson dynasty
- The natural history contribution of keepers in the lighthouse story
- Provides a survey of bird sightings and migration by lighthouse keepers at the end of the 19th century

This beautiful and profusely illustrated book will be of huge appeal to ornithologists, pharologists and everyone with an interest in the natural world and maritime history.

ISBN 978-184995-154-8 hardback £30

Southern Lights

The Scottish Contribution to New Zealand's Lighthouses

- Reveals the contribution made by Scottish lighthouse technology when New Zealand governments built 38 major lighthouses during the period 1859–1941
- Provides a comprehensive and well-researched account of all New Zealand lighthouses including details of the Scottish innovations
- Liberally illustrated, it also features interesting and quirky aspects of the Scottish contribution

ISBN 978-184995-156-2 £18.99